SAS Publishing

Cody's Data Cleaning Techniques Using SAS® Software

Ron Cody

The Power to Know®

The correct bibliographic citation for this manual is as follows: Cody, Ron. 1999. *Cody's Data Cleaning Techniques Using SAS® Software,* Cary, NC: SAS Institute Inc.

Cody's Data Cleaning Techniques Using SAS® Software

Copyright © 1999, SAS Institute Inc., Cary, NC, USA

ISBN 1-58025-600-7

SAS Institute Inc., SAS Campus Drive, Cary, North Carolina 27513.

1st printing, December 1999
2nd printing, February 2002
3rd printing, August 2003
4th printing, March 2005

SAS Publishing provides a complete selection of books and electronic products to help customers use SAS software to its fullest potential. For more information about our e-books, e-learning products, CDs, and hard-copy books, visit the SAS Publishing Web site at **support.sas.com/pubs** or call 1-800-727-3228.

SAS® and all other SAS Institute Inc. product or service names are registered trademarks or trademarks of SAS Institute Inc. in the USA and other countries. ® indicates USA registration.

Other brand and product names are registered trademarks or trademarks of their respective companies.

Table of Contents

1   Checking Values of Character Variables

2   Checking Values of Numeric Variables

3   Checking for Missing Values

4   Working with Dates

5 Looking for Duplicates and "n" Observations per Subject

6 Working with Multiple Files

7 Double Entry and Verification (PROC COMPARE)

8 Some SQL Solutions to Data Cleaning

9 Using Validation Data Sets

Appendix Listing of Raw Data Files and SAS Programs

Index

List of Programs

1 Checking Values of Character Variables

2 Checking Values of Numeric Variables

3 Checking for Missing Values

4 Working with Dates

5 Looking for Duplicates and "n" Observations per Subject

6 Working with Multiple Files

7 Double Entry and Verification (PROC COMPARE)

8 Some SQL Solutions to Data Cleaning

9 Using Validation Data Sets

Introduction

What is data cleaning? In this book, we define data cleaning to include:

- Making sure that the raw data were accurately entered into a computer readable file.
- Checking that character variables contain only valid values.
- Checking that numeric values are within predetermined ranges.
- Checking if there are missing values for variables where complete data is necessary.
- Checking for and eliminating duplicate data entries.
- Checking for uniqueness of certain values, such as patient ID's.
- Checking for invalid date values.
- Checking that an ID number is present in each of "n" files.
- Verifying that more complex multi-file rules have been followed. For example, if an adverse event of type X occurs in one data set, you expect an observation with the same ID number in another data set. In addition, the date of this observation must be after the adverse event and before the end of the trial.

This book provides many programming examples to accomplish the tasks listed above. In many cases, a given problem is solved in several ways. For example, numeric outliers are detected in a DATA step by using formats and informats, by using SAS procedures, and by SQL queries, which are presented together in Chapter 8. Throughout the book, there are useful macros that you may want to add to your collection of data cleaning tools. However, even if you are not experienced with SAS macros, most of the macros that are presented are first presented in non-macro form, so you should still be able to understand the programming concepts that are presented.

But, there is another purpose for this book. It provides instruction on intermediate and advanced SAS programming techniques. One of the reasons for providing multiple solutions to data cleaning problems is to demonstrate specific features of SAS programming. The more complex programs and macros in this book are described in detail.

It is impossible to provide an example of every data cleaning task. Indeed, some studies require custom programming. For those cases, the tools that are developed in this book can be the jumping-off point for more complex programs.

Many applications that require accurate data entry use customized, and sometimes very expensive, data entry and verification programs. A chapter on PROC COMPARE shows how SAS software can be used in a double-entry data verification process.

Chapter 9 describes the use of validation data sets. In a step-by-step process, programs and macros are developed that can read all of the rules for character and numeric variables from a raw data file (called a validation data file) and produce a validation data set and an exception report. The use of integrity constraints, new with Version 7 SAS software, is also discussed.

Although all of the programs in this book were tested by using either Version 7 or Version 8 SAS software, most of the programs should run under Release 6.12, perhaps with some minor changes (such as shortening variable names). However, the integrity constraints discussed in Chapter 9 require using Version 7 or later.

I have enjoyed writing this book. Writing any book is a learning experience and this book is no exception. I hope that most of the egregious errors have been eliminated. If any remain, I take full responsibility for them. Every program in the text has been run against sample data. However, as experience will tell, no program is foolproof.

Acknowledgments

Well, here I am writing the acknowledgments. That means the book is almost finished and I can relax, read a book, ride my bike, or play the piano. Some other people can relax too, and some still have some work to do. My reviewers, Mike Zdeb, Kevin Hobbs, Cynthia Zender, Kent Reeve, George Berg, John Laing, and Grant Cooper finished their work some time ago. I am grateful for their careful reviews, and especially for finding some very subtle errors (and some not so subtle errors).

The editing and production staff are still working. Judy Whatley, although she is listed as the acquisitions editor, has played a much larger role, overseeing the book throughout the production process. Judy, it's been a real pleasure working with you. Thanks so much for your upbeat attitude (even through major floods in North Carolina). Other SAS Institute folks who played an important role in bringing this book to print are copyeditor Josephine Pope, production specialist Mary Rios, marketing analyst Patricia Urquhart, cartoonist Mike Pezzoni, and cover designer Cate Parrish. However, since I haven't seen the cover yet, I may want to withhold my appreciation to Cate! Just kidding.

Ron Cody
Winter 2000

1 Checking Values of Character Variables

Introduction

There are some basic operations that need to be routinely performed when dealing with character data values. You may have a character variable that can take on only certain allowable values, such as 'M' and 'F' for gender. You may also have a character variable that can take on numerous values but the values must fit a certain form, such as a single letter followed by two or three digits. This chapter shows you several ways that you can use SAS software to perform validity checks on character variables.

Using PROC FREQ to List Values

This section demonstrates how to use PROC FREQ to check for invalid values of a character variable. In order to test the programs you develop, use the raw data file PATIENTS.TXT, listed in the Appendix. You can use this data file and, in later sections, a SAS data set created from this raw data file for many of the examples in this text.

Description of the Raw Data File PATIENTS.TXT

The raw data file PATIENTS.TXT contains both character and numeric variables from a typical clinical trial. A number of data errors were included in the file so that you can test the data cleaning programs that are developed in this text. The programs in this book assume that the file PATIENTS.TXT is located in a directory (folder) called C:\CLEANING. This is the directory that is used throughout this text as the location for data files, SAS data sets, SAS programs, and SAS macros. You can modify the INFILE and LIBNAME statements to fit your own operating environment.

Here is the layout for the data file PATIENTS.TXT.

Variable Name	Description	Starting Column	Length	Variable Type	Valid Values
PATNO	Patient Number	1	3	Character	Numerals only
GENDER	Gender	4	1	Character	'M' or 'F'
VISIT	Visit Date	5	10	MMDDYY10.	Any valid date
HR	Heart Rate	15	3	Numeric	Between 40 and 100
SBP	Systolic Blood Pressure	18	3	Numeric	Between 80 and 200
DBP	Diastolic Blood Pressure	21	3	Numeric	Between 60 and 120
DX	Diagnosis Code	24	3	Character	1 to 3 digit numeral
AE	Adverse Event	27	1	Character	'0' or '1'

There are several character variables that should have a limited number of valid values. For this exercise, you expect values of GENDER to be 'F' or 'M', values of DX the numerals 1 through 999, and values of AE (adverse events) to be '0' or '1'. A very simple approach to identifying invalid character values in this file is to use PROC FREQ to list all the unique values of these variables. Of course, once invalid values are identified using this technique, other means will have to be employed to locate specific records (or patient numbers) corresponding to the invalid values.

Use the program PATIENTS.SAS (shown next) to create the SAS data set PATIENTS from the raw data file PATIENTS.TXT (which can be downloaded from the SAS Web site or found listed in the Appendix). This program is followed with the appropriate PROC FREQ statements to list the unique values (and their frequencies) for the variables GENDER, DX, and AE.

Program 1-1   Writing a Program to Create the Data Set PATIENTS

```
*------------------------------------------------------------*
|PROGRAM NAME: PATIENTS.SAS IN C:\CLEANING                   |
|PURPOSE: TO CREATE A SAS DATA SET CALLED PATIENTS           |
*-----------------------------------------------------------*;
LIBNAME CLEAN "C:\CLEANING";

DATA CLEAN.PATIENTS;
   INFILE "C:\CLEANING\PATIENTS.TXT" PAD; /* Pad short records
                                            with blanks  */

   INPUT @1  PATNO    $3. @4  GENDER    $1.
         @5  VISIT    MMDDYY10.
         @15 HR       3.
         @18 SBP      3.
         @21 DBP      3.
         @24 DX       $3.
         @27 AE       $1.;

   LABEL PATNO    = "Patient Number"
         GENDER   = "Gender"
         VISIT    = "Visit Date"
         HR       = "Heart Rate"
         SBP      = "Systolic Blood Pressure"
         DBP      = "Diastolic Blood Pressure"
         DX       = "Diagnosis Code"
         AE       = "Adverse Event?";
   FORMAT VISIT MMDDYY10.;
RUN;
```

The DATA step is straightforward. Notice the PAD option in the INFILE statement. This will seem foreign to most mainframe users and is probably no longer necessary on other platforms. The PAD option pads all records (adds blanks to the end of short records) to the default logical record length or a length specified by another INFILE option, LRECL. It prevents the possibility of skipping to the next record (line) of data when a short line is encountered.

Next, you want to use PROC FREQ to list all the unique values for your character variables. To simplify the output from PROC FREQ, use the NOCUM (no cumulative statistics) and NOPERCENT (no percentages) TABLES options because you only want frequency counts for each of the unique character values. (Note, sometimes the percent and cumulative statistics can be useful — the choice is yours.) The PROC statements are shown in Program 1-2.

Program 1-2   Using PROC FREQ to List All the Unique Values for Character Variables

```
PROC FREQ DATA=CLEAN.PATIENTS;
   TITLE "Frequency Counts for Selected Character Variables";
   TABLES GENDER DX AE / NOCUM NOPERCENT;
RUN;
```

Here is the output from running Program 1-2.

```
Frequency Counts for Selected Character Variables

The FREQ Procedure

        Gender

GENDER    Frequency
-------------------
2             1
F            12
M            14
X             1
f             2

Frequency Missing = 1


Diagnosis Code

DX     Frequency
---------------
1          7
2          2
3          3
4          3
5          3
6          1
7          2
X          2

Frequency Missing = 8



Adverse Event?

AE     Frequency
---------------
0         19
1         10
A          1

Frequency Missing = 1
```

Let's focus in on the frequency listing for the variable GENDER. If valid values for GENDER are 'F', 'M', and missing, this output would point out several data errors. The values '2' and 'X' both occur once. Depending on the situation, the lowercase value 'f' may or may not be considered an error. If lowercase values were entered into the file by mistake, but the value (aside from the case) was correct, you could change all lowercase values to uppercase with the UPCASE function. More on that later. The invalid DX code of 'X' and the adverse event of 'A' are also easily identified. At this point, it is necessary to run additional programs to identify the location of these errors. Running PROC FREQ is still a useful first step in identifying errors of these types, and it is also useful as a last step, after the data have been cleaned, to ensure that all the errors have been identified and corrected.

Using a DATA Step to Check for Invalid Values

Your next task is to use a DATA step to identify invalid data values and to determine where they occur in the raw data file (by listing the patient number).

This time, DATA step processing is used to identify invalid character values for selected variables. As before, you will check GENDER, DX, and AE. Several different methods are used to identify these values.

First, you can write a simple DATA step that reports invalid data values by using PUT statements in a DATA _NULL_ step. Here is the program.

Program 1-3   Using a DATA _NULL_ Step to Detect Invalid Character Data

```
DATA _NULL_;
   INFILE "C:\CLEANING\PATIENTS.TXT" PAD;
   FILE PRINT; ***Send output to the Output window;
   TITLE "Listing of Invalid Patient Numbers and Data Values";
   ***Note: We will only input those variables of interest;
   INPUT @1  PATNO    $3.
         @4  GENDER   $1.
         @24 DX       $3.
         @27 AE       $1.;
   ***Check GENDER;
   IF GENDER NOT IN ('F' 'M' ' ') THEN PUT PATNO= GENDER=;
   ***Check DX;
   IF VERIFY(DX,' 0123456789') NE 0 THEN PUT PATNO= DX=;
   ***Check AE;
   IF AE NOT IN ('0' '1' ' ') THEN PUT PATNO= AE=;
RUN;
```

Before discussing the output, let's spend a moment looking over the program. First, notice the use of the DATA _NULL_ statement. Because the only purpose of this program is to identify invalid data values, there is no need to create a SAS data set. The FILE PRINT statement causes the results of any subsequent PUT statements to be sent to the Output window (or output device). Without this statement, the results of the PUT statements would be sent to the SAS Log. GENDER and AE are checked by using the IN operator. The statement

```
IF X IN ('A' 'B' 'C') THEN . . .;
```

is equivalent to

```
IF X = 'A' OR X = 'B' OR X = 'C' THEN . . .;
```

That is, if X is equal to any of the values in the list following the IN operator, the expression is evaluated as true. You want an error message printed when the value of GENDER is not one of the acceptable values ('F', 'M', or missing). Therefore, place a NOT in front of the whole expression, triggering the error report for invalid values of GENDER or AE. You can separate the values in the list by spaces or commas.

There are several alternative ways that the gender checking statement can be written. The method above uses the IN operator.

A straightforward alternative to the IN operator is

```
IF NOT (GENDER EQ 'F' OR GENDER EQ 'M' OR GENDER = ' ') THEN
PUT PATNO= GENDER=;
```

Another possibility is

```
IF GENDER NE 'F' AND GENDER NE 'M' AND GENDER NE ' ' THEN
PUT PATNO= GENDER=;
```

While all of these statements checking for GENDER and AE produce the same result, the IN operator is probably the easiest to write, especially if there are a large number of possible values to check. Always be sure to consider whether you want to identify missing values as invalid or not. In the statements above, you are allowing missing values as valid codes. If you want to flag missing values as errors, do not include a missing value in the list of valid codes.

If you want to allow lowercase M's and F's as valid values, you can add the single line

```
GENDER = UPCASE(GENDER);
```

immediately before the line that checks for invalid gender codes. As you can probably guess, the UPCASE function changes all lowercase letters to uppercase letters.

A statement similar to the gender checking statement is used to test the adverse events.

There are so many valid values for DX (any numeral from 1 to 999), that the approach you used for GENDER and AE would be inefficient (and wear you out typing) if you used it to check for invalid DX codes. The VERIFY function is one of the many possible ways you can check to see if there is a value other than the numerals 0 to 9 or blank as a DX value. The VERIFY function has the following form:

```
VERIFY(character_variable,verify_string)
```

where the verify string is either a character variable or a series of character values placed in single or double quotes. The VERIFY function returns the first position in the character_variable that contains a character that is not in the verify_string. If the character_variable does not contain any invalid values, the VERIFY function returns a 0. To make this clearer, let's look at the following statement that uses the VERIFY function to check for invalid GENDER values:

```
IF VERIFY (GENDER,'FM ') NE 0 THEN PUT PATNO= GENDER=;
```

Notice that you included a blank in the verify_string so that missing values will be considered valid. If GENDER has a value other than an 'F', 'M', or missing, the VERIFY function returns the position of the invalid character in the string. But, because the length of GENDER is 1, any invalid value for GENDER returns a 1.

You are now ready to understand the VERIFY function that checked for invalid DX codes. The verify string contained a blank plus the characters (numerals) 0 through 9. Thus, if the DX code contains any character other than a blank or a 0 through 9, it returns the position of this offending character, which would have to be a 1, 2, or 3 (DX is three bytes in length), and the error message would be printed.

Although the function

```
VERIFY(DX,' 0123456789')
```

returns a 0 if there are no invalid characters in the DX code, it should be pointed out that DX codes with embedded blanks will not be identified as invalid with this statement. If you want to ensure that only the character representations of the numbers 1 to 999 are considered valid, the following statements can be used:

```
X_DX = INPUT(DX,3.);
IF X_DX EQ . AND DX NE ' ' THEN PUT PATNO= DX=;
```

The first line above creates a numeric variable (X_DX) from the character DX value. The INPUT function can be thought of in a similar manner to an INPUT statement. It says to pretend you are reading a variable (DX) from a data file according to the INFORMAT 3. except you are actually "reading" the value from a character variable. The result of this process is to be assigned to the variable X_DX. In other words, the INPUT function performs a character-to-numeric conversion. If there is an invalid DX code (containing a letter or embedded blank, for example), the INPUT function sends an error message to the SAS Log and returns a missing value. In the second line, you test if the numeric equivalent of the DX code is missing and the original DX is not missing, putting out an error message when this condition is true. (Note, because the original character value was three bytes, you don't have to test if X_DX is greater than 999, because this is the largest number you can write with three digits.) Any invalid DX code will then cause the error message to be printed.

For really compulsive programmers (like the author), there is one final problem with the above approach. A value such as 1.3 would not result in an error message because the number 1.3 is between 1 and 999. There are several ways around this problem. One way is to use the TRANSLATE function to substitute an invalid character for the decimal point before you perform the character-to-numeric conversion.

```
X_DX = INPUT(TRANSLATE(DX,'X','.'),3.);
```

The TRANSLATE function above will convert periods (or decimal points) to X's. If DX originally contained a decimal point, the value of X_DX would be a missing value. In general, the syntax of the TRANSLATE function is

```
TRANSLATE(char_variable,to_string,from_string)
```

where each character in the from_string is translated to the corresponding character in the to_string. For example, to translate the numerals 1 through 5 to the letters A through E for a variable called SCORE, you would write

```
NEW_VAR = TRANSLATE(SCORE,'ABCDE','12345');
```

Another interesting approach is to test to see if the value of X_DX is not an integer. The MOD function is an effective way to do this. If any number modulus 1 is not 0 (the remainder after you divide the number by 1), the number is not an integer. The SAS code using this method is

```
X_DX = INPUT(DX,3.);
IF (X_DX EQ . OR MOD(X_DX,1) NE 0) AND
   DX NE ' ' THEN PUT PATNO= DX=;
```

Here is another point. If you want to avoid filling up your SAS Log with error messages resulting from invalid arguments to the INPUT function, you can use the double question mark (??) modifier before the informat to tell the program to ignore these errors and not to report the errors to the SAS Log. The INPUT function would then look like this:

```
X_DX = INPUT(DX,?? 3.);
```

The ?? informat modifier can also be used with the INPUT statement. Here is the output from running Program 1-3.

```
Listing of Invalid Patient Numbers and Data Values

PATNO=002   DX=X
PATNO=003   GENDER=X
PATNO=004   AE=A
PATNO=010   GENDER=f
PATNO=013   GENDER=2
PATNO=002   DX=X
PATNO=023   GENDER=f
```

Note that patient 002 appears twice in this output. This occurs because there is a duplicate observation for patient 002 (in addition to several other purposely included errors), so that the data set can be used for examples later in this book, such as the detection of duplicate ID's and duplicate observations.

Suppose you want to check for valid patient numbers (PATNO) in a similar manner. However, you want to flag missing values as errors (every patient must have a valid ID). The following statements:

```
ID = INPUT(TRANSLATE(PATNO,'X','.'),?? 3.);
IF ID LT 1 THEN PUT "Invalid ID for PATNO=" PATNO;
```

will work in the same way as your check for invalid DX codes except that missing values will now be listed as errors.

Using PROC PRINT with a WHERE Statement to List Invalid Values

There are several alternative ways to identify the ID's containing invalid data. As with most of the topics in this book, you will see several ways of accomplishing the same task. Why? One reason is that some techniques are better suited to an application. Another reason is to teach some additional SAS programming techniques. Finally, under different circumstances, some techniques may be more efficient than others.

One very easy alternative way to list the subjects with invalid data is to use PROC PRINT followed by a WHERE statement. Just as you used an IF statement in a DATA step in the previous section, you can use a WHERE statement in a similar manner with PROC PRINT and avoid having to write a DATA step altogether. For example, to list the ID's with invalid GENDER values, you could write a program like the one shown in Program 1-4.

Program 1-4   Using PROC PRINT to List Invalid Character Values

```
PROC PRINT DATA=CLEAN.PATIENTS;
   TITLE "LISTING OF INVALID GENDER VALUES";
   WHERE GENDER NOT IN ('M' 'F' ' ');
   ID PATNO;
   VAR GENDER;
RUN;
```

It's easy to forget that WHERE statements can be used within SAS procedures. SAS programmers that have been at it for a long time (like the author) often write a short DATA step first and use PUT statements or create a temporary SAS data set and follow it with a PROC PRINT. The program above is both shorter and more efficient than a DATA step followed by a PROC PRINT. DATA _NULL_ steps, however, tend to be fairly efficient and are a reasonable alternative as well as the more flexible approach.

The output from Program 1-4 follows.

```
LISTING OF INVALID GENDER VALUES

PATNO    GENDER

 003       X
 010       f
 013       2
 023       f
```

This program can be extended to list invalid values for all the character variables. You simply add the other invalid conditions to the WHERE statement as shown in Program 1-5.

Program 1-5   Using PROC PRINT to List Invalid Character Data for Several Variables

```
PROC PRINT DATA=CLEAN.PATIENTS;
   TITLE "LISTING OF INVALID CHARACTER VALUES";
   WHERE GENDER NOT IN ('M' 'F' ' ')           OR
         VERIFY(DX,' 0123456789') NE 0          OR
         AE NOT IN ('0' '1' ' ');
   ID PATNO;
   VAR GENDER DX AE;
RUN;
```

The resulting output is shown next.

```
LISTING OF INVALID CHARACTER VALUES

PATNO    GENDER    DX    AE

 002       F       X      0
 003       X       3      1
 004       F       5      A
 010       f       1      0
 013       2       1
 002       F       X      0
 023       f              0
```

Notice that this output is not as informative as the one produced by the DATA _NULL_ step in Program 1-3. It lists all the patient numbers, genders, DX codes, and adverse events even when only one of the variables has an error (patient 002 for example). So, there is a trade-off — the simpler program produces slightly less desirable output. We could get philosophical and extend this concept to life in general, but, that's for some other book.

You can also substitute any of the more complicated logical expressions from the previous section into this WHERE statement if you wish. For example, to perform a more careful check on DX codes, you could modify the WHERE statement as shown here.

```
PROC PRINT DATA=CLEAN.PATIENTS;
   TITLE "LISTING OF INVALID CHARACTER VALUES";
   WHERE GENDER NOT IN ('M' 'F' ' ')                        OR
      (INPUT(DX,3.) EQ . OR MOD(INPUT(DX,3.),1) NE 0)    AND
      DX NE ' '                                          OR
      AE NOT IN ('0' '1' ' ');
   ID PATNO;
   VAR GENDER DX AE;
RUN;
```

Using Formats to Check for Invalid Values

Another way to check for invalid values of a character variable from raw data is to use user-defined formats. There are several possibilities here. One, you can create a format that leaves all valid character values as is and formats all invalid values to a single error code. Let's start out with a program that simply assigns formats to the character variables and uses PROC FREQ to list the number of valid and invalid codes. Following that, you will extend the program by using a DATA step to identify which ID's have invalid values. Program 1-6 uses formats to convert all invalid data values to a single value.

Program 1-6   Using a User-Defined Format and PROC FREQ to List Invalid Data Values

```
PROC FORMAT;
   VALUE $GENDER 'F','M' = 'Valid'
                  ' '    = 'Missing'
                  OTHER  = 'Miscoded';
   VALUE $DX '001' - '999' = 'Valid'  /* See important note below */
             ' '        = 'Missing'
             OTHER      = 'Miscoded';

   VALUE $AE '0','1' = 'Valid'
             ' '     = 'Missing'
             OTHER   = 'Miscoded';
RUN;

PROC FREQ DATA=CLEAN.PATIENTS;
   TITLE "Using Formats to Identify Invalid Values";
   FORMAT GENDER $GENDER.
          DX     $DX.
          AE     $AE.;
   TABLES GENDER DX AE / NOCUM NOPERCENT MISSING;
RUN;
```

For the variables GENDER and AE, which have specific valid values, you list each of the valid values in the range to the left of the equal sign in the VALUE statement. Format each of these values with the value 'Valid'. For the $DX format, you specify a range of values on the left side of the equal sign.

Important Note: It should be pointed out here, that the range '001' - '999' will behave differently on Windows and UNIX platforms compared to MVS and CMS platforms. You may want to test several values on your platform to be sure the program is performing as you intend. For example, the value '0A1' will be considered 'Valid' on a Windows or a UNIX platform and 'Invalid' on MVS or CMS (as pointed out by two of my reviewers, John Laing and Mike Zdeb). You may want to test for alphabetic values for DX in a short DATA step, prior to running Program 1-6.

You may choose to lump the missing value with the valid values if that is appropriate, or you may want to keep track of missing values separately as was done here. Finally, any value other than the valid values or a missing value will be formatted as 'Miscoded'. All that is left is to run PROC FREQ to count the number of 'Valid', 'Missing', and 'Miscoded' values. The TABLES option MISSING causes the missing values to be listed in the body of the PROC FREQ output. Here is the output from PROC FREQ.

```
Using Formats to Identify Invalid Values
The FREQ Procedure

        Gender
GENDER        Frequency
--------------------
Missing            1
Miscoded           4
Valid             26

Diagnosis Code

DX            Frequency
--------------------
Missing            8
Valid             21
Miscoded           2

Adverse Event?

AE            Frequency
--------------------
Missing            1
Valid             29
Miscoded           1
```

This output isn't particularly useful. It doesn't tell you which observations (patient numbers) contain missing or invalid values. Let's modify the program by adding a DATA step, so that ID's with invalid character values are listed.

Program 1-7   Using a User-Defined Format and a DATA Step to List Invalid Data Values

```
PROC FORMAT;
   VALUE $GENDER 'F','M' = 'Valid'
                 ' '      = 'Missing'
              OTHER   = 'Miscoded';
   VALUE $DX '001' - '999' = 'Valid'
             ' '            = 'Missing'
              OTHER     = 'Miscoded';
   VALUE $AE '0','1' = 'Valid'
             ' '      = 'Missing'
            OTHER  = 'Miscoded';
RUN;
```

```
DATA _NULL_;
    INFILE "C:\CLEANING\PATIENTS.TXT" PAD;
    FILE PRINT; ***Send output to the Output window;
    TITLE "Listing of Invalid Patient Numbers and Data Values";
    ***Note: We will only input those variables of interest;
    INPUT @1  PATNO    $3.
          @4  GENDER   $1.
          @24 DX       $3.
          @27 AE       $1.;

    IF PUT(GENDER,$GENDER.) = 'Miscoded' THEN PUT PATNO= GENDER=;
    IF PUT(DX,$DX.) = 'Miscoded' THEN PUT PATNO= DX=;
    IF PUT(AE,$AE.) = 'Miscoded' THEN PUT PATNO= AE=;
RUN;
```

The "heart" of this program is the PUT function. To review, the PUT function is similar to the INPUT function. It takes the following form:

```
character_variable = PUT(variable,format)
```

where character_variable is a character variable that contains the value of the variable listed as the first argument to the function, formatted by the format listed as the second argument to the function. The result of a PUT function is always a character variable and the function is frequently used to perform numeric-to-character conversions. In Program 1-7, the first argument of the PUT function is a character variable, and the result of the PUT function for any invalid data values would be the value 'Miscoded'.

Here is the output from Program 1-7.

```
Listing of Invalid Patient Numbers and Data Values

PATNO=002   DX=X
PATNO=003   GENDER=X
PATNO=004   AE=A
PATNO=010   GENDER=f
PATNO=013   GENDER=2
PATNO=002   DX=X
PATNO=023   GENDER=f
```

Using Informats to Check for Invalid Values

PROC FORMAT is also used to create informats. Remember that formats are used to control how variables look in output or how they are classified by such procedures as PROC FREQ. Informats modify the value of variables as they are read from the raw data, or they can be used with an INPUT function to create new variables in the DATA step. User-defined informats are created in much the same way as user-defined formats. Instead of a VALUE statement that creates formats, an INVALUE statement is used to create informats. The only difference between the two is that informat names can only be seven characters in length. (Note: For those curious readers, the reason is that informats and formats are both stored in the same catalog and an "@" is placed before informats to distinguish them from formats.) The following is a program that changes invalid values for GENDER and AE to missing values by using a user-defined informat.

Program 1-8   Using a User-Defined Informat to Set Invalid Data Values to Missing

```
*-----------------------------------------------------------------*
|  PROGRAM NAME: INFORM1.SAS IN C:\CLEANING                       |
|  PURPOSE: TO CREATE A SAS DATA SET CALLED PATIENTS2             |
|           AND SET ANY INVALID VALUES FOR GENDER AND AE TO       |
|           MISSING, USING A USER-DEFINED INFORMAT                |
*-----------------------------------------------------------------*;
LIBNAME CLEAN "C:\CLEANING";

PROC FORMAT;
    INVALUE $GEN      'F','M' = _SAME_
                      OTHER   = ' ';
    INVALUE $AE       '0','1' = _SAME_
                      OTHER   = ' ';
RUN;

DATA CLEAN.PATIENTS2;
    INFILE "C:\CLEANING\PATIENTS.TXT" PAD;
    INPUT @1  PATNO    $3.
          @4  GENDER   $GEN1.
          @27 AE       $AE1.;

    LABEL PATNO   = "Patient Number"
          GENDER  = "Gender"
          DX      = "Diagnosis Code"
          AE      = "Adverse Event?";
RUN;
```

```
PROC PRINT DATA=CLEAN.PATIENTS2;
   TITLE "Listing of Data Set PATIENTS2";
   VAR PATNO GENDER AE;
RUN;
```

Notice the INVALUE statements in the PROC FORMAT above. The key word _SAME_ is a SAS reserved value that does what its name implies — it leaves any of the values listed in the range specification unchanged. The key word OTHER in the subsequent line refers to any values not matching one of the previous ranges. Notice also, that the informats in the INPUT statement use the user-defined informat name followed by the number of columns to be read, the same method that is used with predefined SAS informats.

Output from the PROC PRINT is shown next.

```
Listing of Data Set PATIENTS2

 Obs     PATNO     GENDER      AE

   1      001         M         0
   2      002         F         0
   3      003                   1
   4      004         F
   5      XX5         M         0
   6      006                   1
   7      007         M         0
   8                  M         0
   9      008         F         0
  10      009         M         1
  11      010                   0
  12      011         M         1
  13      012         M         0
  14      013
  15      014         M         1
  16      002         F         0
  17      003         M         0
  18      015         F         1
  19      017         F         0
  20      019         M         0
  21      123         M         0
  22      321         F         1
  23      020         F         0
  24      022         M         1
  25      023                   0
  26      024         F         0
  27      025         M         1
  28      027         F         0
  29      028         F         0
  30      029         M         1
  31      006         F         0
```

Notice that invalid values for GENDER and AE are now missing values, including the two lowercase 'f's (patient numbers 010 and 023).

Let's add one more feature to this program. By using the keyword UPCASE in the informat specification, you can automatically convert the values being read to uppercase before the ranges are checked. Here are the PROC FORMAT statements, rewritten to use this option.

```
PROC FORMAT;
   INVALUE $GEN (UPCASE)   'F' = 'F'
                           'M' = 'M'
                         OTHER = ' ' ;
   INVALUE $AE '0','1' = _SAME_
            OTHER  = ' ' ;
RUN;
```

The UPCASE option is placed in parenthesis following the informat name. Notice some other changes as well. You cannot use the keyword _SAME_ anymore because the value is changed to uppercase for comparison purposes, but the _SAME_ specification would leave the original lowercase value unchanged. By specifying each value individually, the lowercase 'f' (the only lowercase GENDER value) would match the range 'F' and be assigned the value of an uppercase 'F'.

The output of this data set is identical to the output for Program 1-8 except the value of GENDER for patients 010 and 023 are an uppercase 'F'.

If you want to preserve the original value of the variable, you can use a user-defined informat with an INPUT function instead of an INPUT statement. You can use this method to check a raw data file or a SAS data set. Program 1-9 reads the SAS data set CLEAN.PATIENTS and uses user-defined informats to detect errors.

Program 1-9   Using a User-Defined Informat with the INPUT Function

```
PROC FORMAT;
   INVALUE $GENDER 'F','M' = _SAME_
                   OTHER  = 'ERROR';
   INVALUE $AE     '0','1' = _SAME_
                   OTHER   = 'ERROR';
RUN;

DATA _NULL_;
   FILE PRINT;
   SET CLEAN.PATIENTS;
   IF INPUT (GENDER,$GENDER.) = 'ERROR' THEN
      PUT @1 "Error for Gender for Patient:" PATNO" Value is " GENDER;
   IF INPUT (AE,$AE.) = 'ERROR' THEN
      PUT @1 "Error for AE for Patient:" PATNO" Value is " AE;
RUN;
```

The advantage of this program over Program 1-8 is that the original values of the variables are not lost.

2 Checking Values of Numeric Variables

Introduction

The techniques for checking for invalid numeric data are quite different from the techniques that you saw in the last chapter for checking character data. Although there are usually many different values a numeric variable can take on, there are several techniques that you can use to help identify data errors. One simple technique is to examine some of the largest and smallest data values for each numeric variable. If you see values such as 12 or 1200 for a systolic blood pressure (usually between 80 and 200 in healthy adults), you can be quite certain that an error was made, either in entering the data values or on the original data collection form.

There are also some internal consistency methods that can be used to identify possible data errors. If you see that most of the data values fall within a certain range of values, then any values that fall far enough outside that range may be data errors. This chapter develops programs based on these ideas.

Using PROC MEANS, PROC TABULATE, and PROC UNIVARIATE to Look for Outliers

One of the simplest ways to check for invalid numeric values is to run either PROC MEANS or PROC UNIVARIATE. By default, PROC MEANS lists the minimum and maximum values, along with the n, mean, and standard deviation. PROC UNIVARIATE is somewhat more useful in detecting invalid values, because it provides you with a listing of the five highest and five lowest values, along with graphical output (stem-and-leaf plots and box plots). Let's first look at how you can use PROC MEANS for very simple checking of numeric variables. The program below checks the three numeric variables, heart rate (HR), systolic blood pressure (SBP), and diastolic blood pressure (DBP), in the PATIENTS data set.

Program 2-1   Using PROC MEANS to Detect Invalid and Missing Values

```
LIBNAME CLEAN "C:\CLEANING";
PROC MEANS DATA=CLEAN.PATIENTS N NMISS MIN MAX MAXDEC=3;
   TITLE "Checking Numeric Variables in the PATIENTS Data Set";
   VAR HR SBP DBP;
RUN;
```

Let's choose the options N, NMISS, MIN, MAX, and MAXDEC=3 for this procedure. The N and NMISS options report the number of nonmissing and missing observations for each variable, respectively. The MIN and MAX options list the smallest and largest nonmissing values for each variable. The MAXDEC=3 option is used so that the minimum and maximum values will be printed to three decimal places. Because HR, SBP, and DBP are supposed to be integers, you might have thought to set the MAXDEC option to 0. However, you might want to catch any data errors where a decimal point was entered by mistake.

Here is the output from Program 2-1.

```
Checking Numeric Variables in the PATIENTS Data Set

The MEANS Procedure

                                               N
Variable  Label                      N        Miss     Minimum      Maximum
---------------------------------------------------------------------------
HR        Heart Rate                 28       3         10.000      900.000
SBP       Systolic Blood Pressure    27       4         20.000      400.000
DBP       Diastolic Blood Pressure   28       3          8.000      200.000
---------------------------------------------------------------------------
```

This output is not particularly useful. It does show the number of nonmissing and missing observations along with the highest and lowest values. Inspection of the minimum and maximum values for all three variables shows that there are probably some data errors in the PATIENTS data set. If you want a slightly prettier output, you can use PROC TABULATE to accomplish the same task. For an excellent reference on PROC TABULATE, let me suggest a book written by Lauren E. Haworth, called *PROC TABULATE by Example*, published by SAS Institute, Cary, NC, as part of their Books by Users series.

Here is the equivalent PROC TABULATE program, followed by the output. (Assume that the libref CLEAN has been previously defined in this program and in any future programs where it is not included in the program.)

Program 2-2   Using PROC TABULATE to Display Descriptive Data

```
PROC TABULATE DATA=CLEAN.PATIENTS FORMAT=7.3;   ❶
   TITLE "Statistics for Numeric Variables";
   VAR HR SBP DBP;   ❷
   TABLES HR SBP DBP,
          N*F=7.0 NMISS*F=7.0 MEAN MIN MAX / RTSPACE=18;   ❸
   KEYLABEL N     = 'Number'   ❹
            NMISS = 'Missing'
            MEAN  = 'Mean'
            MIN   = 'Lowest'
            MAX   = 'Highest';
RUN;
```

The FORMAT option ❶ tells the procedure to use the numeric format 7.3 (a field width of 7 with 3 places to the right of the decimal point) for all the output, unless otherwise specified. The analysis variables HR, SBP, and DBP are listed in a VAR statement ❷. Let's place these variables on the row dimension and the statistics along the column dimension. The TABLE option RTSPACE=18 ❸ allows for 18 spaces for all row labels, including the spaces for the lines forming the table. In addition, the format 7.0 is to be used for N and NMISS in the table. Finally, the KEYLABEL statement ❹ replaces the keywords for the selected statistics with more meaningful labels. Below is the output from PROC TABULATE.

Statistics for Numeric Variables

	Number	Missing	Mean	Lowest	Highest
Heart Rate	28	3	107.393	10.000	900.000
Systolic Blood Pressure	27	4	144.519	20.000	400.000
Diastolic Blood Pressure	28	5	88.071	8.000	200.000

A more useful procedure might be PROC UNIVARIATE. Running this procedure for your numeric variables yields much more information.

Program 2-3   Using PROC UNIVARIATE to Look for Outliers

```
PROC UNIVARIATE DATA=CLEAN.PATIENTS PLOT;
   TITLE "Using PROC UNIVARIATE to Look for Outliers";
   VAR HR SBP DBP;
RUN;
```

The procedure option PLOT provides you with several graphical displays of the data; a stem-and-leaf plot, a box plot, and a normal probability plot. Output from this procedure is shown next. (**Note:** To save some space, the PROC UNIVARIATE output for the variable SBP has been omitted)

```
Using PROC UNIVARIATE to Look for Outliers

The UNIVARIATE Procedure
Variable:  HR   (Heart Rate)

                            Moments

N                        28    Sum Weights                  28
Mean              107.392857   Sum Observations           3007
Std Deviation     161.086436   Variance             25948.8399
Skewness          4.73965876   Kurtosis             23.7861582
Uncorrected SS       1023549   Corrected SS         700618.679
Coeff Variation   149.997347   Std Error Mean        30.442475

                  Basic Statistical Measures

       Location                     Variability

Mean      107.3929    Std Deviation         161.08644
Median     74.0000    Variance                  25949
Mode       68.0000    Range               890.00000
                      Interquartile Range  27.00000

              Tests for Location: Mu0=0

Test              -Statistic-      -----p Value------

Student's t    t  3.527731    Pr > |t|      0.0015
Sign           M        14    Pr >= |M|     <.0001
Signed Rank    S       203    Pr >= |S|     <.0001

Quantiles (Definition 5)

Quantile        Estimate

100% Max            900
99%                 900
95%                 210
90%                 208
75% Q3               87
50% Median           74
25% Q1               60
10%                  22
5%                   22
1%                   10
0% Min               10

                                                   Continued
```

```
Using PROC UNIVARIATE to Look for Outliers

The UNIVARIATE Procedure
Variable:  HR  (Heart Rate)

          Extreme Observations

----Lowest----        ----Highest---

Value      Obs      Value       Obs

   10       23         90         8
   22       25        101         4
   22       15        208        19
   48       24        210         9
   58       20        900        22


              Missing Values

                   -----Percent Of-----
Missing                            Missing
   Value       Count     All Obs      Obs

     .            3        9.68     100.00


   Stem Leaf                        #  Boxplot
      9 0                           1     *
      8
      7
      6
      5
      4
      3
      2 11                          2     *
      1 0                           1     +
      0 1225666677777778888889999  24  +--0--+
        ----+----+----+----+----
      Multiply Stem.Leaf by 10**+2


Using PROC UNIVARIATE to Look for Outliers

The UNIVARIATE Procedure
Variable:  HR  (Heart Rate)

                   Normal Probability Plot
    950+                                      *
       |
       |
    650+                                         +
       |                                   ++++++
    350+                              +++++++
       |                         +++++ *   *
       |                    ++++++    *
     50+     *    *    **  *+**+*****  *  **   *
       +----+----+----+----+----+----+----+----+----+----+
           -2        -1         0        +1        +2
```

Continued

```
Using PROC UNIVARIATE to Look for Outliers

The UNIVARIATE Procedure
Variable: DBP  (Diastolic Blood Pressure)

                            Moments

N                         28      Sum Weights                   28
Mean               88.0714286      Sum Observations            2466
Std Deviation      37.2915724      Variance              1390.66138
Skewness           1.06190956      Kurtosis              3.67139184
Uncorrected SS         254732      Corrected SS          37547.8571
Coeff Variation     42.342418      Std Error Mean        7.04744476


                   Basic Statistical Measures

        Location                        Variability

Mean       88.07143      Std Deviation            37.29157
Median     81.00000      Variance                     1391
Mode       78.00000      Range                   192.00000
                         Interquartile Range      26.00000

NOTE: The mode displayed is the smallest of 2 modes with a
      count of 3.


              Tests for Location: Mu0=0

Test               -Statistic-      -----p Value------

Student's t     t  12.49693      Pr > |t|    <.0001
Sign            M        14      Pr >= |M|   <.0001
Signed Rank     S       203      Pr >= |S|   <.0001


         Quantiles (Definition 5)

Quantile      Estimate

100% Max          200
99%               200
95%               180
90%               120
75% Q3            100
50% Median         81
25% Q1             74
10%                64
5%                 20
1%                  8
0% Min              8
```
                                                            Continued

```
Using PROC UNIVARIATE to Look for Outliers

The UNIVARIATE Procedure
Variable:  DBP  (Diastolic Blood Pressure)
          Extreme Observations

----Lowest----        ----Highest---

Value      Obs         Value      Obs

    8       23          106        28
   20       12          120         4
   64       14          120        11
   68       27          180        10
   68        6          200        22


            Missing Values


                    -----Percent Of-----
Missing                           Missing
  Value       Count     All Obs       Obs

      .           3        9.68    100.00

   Stem Leaf                    #  Boxplot
     20 0                       1    *
     18 0                       1    *
     16
     14
     12 00                      2    |
     10 0026                    4  +-----+
      8 000244800               9  *--+--*
      6 488044888               9  +-----+
      4
      2 0                       1    0
      0 8                       1    0
        ----+----+----+----+
     Multiply Stem.Leaf by 10**+1


Using PROC UNIVARIATE to Look for Outliers

The UNIVARIATE Procedure
Variable:  DBP  (Diastolic Blood Pressure)
                 Normal Probability Plot
    210+                                      *
       |                                *
       |                                   +++++
       |                              +++++
       |                          ++*+*
    110+                    ++*** *
       |               ****+**
       |        * *  **+*+*
       |         +++++
       |      ++*+++
     10+  +++*+
        +----+----+----+----+----+----+----+----+----+----+
          -2        -1         0        +1        +2
```

You certainly get lots of information from PROC UNIVARIATE, perhaps too much information. Starting off, you see some descriptive univariate statistics (hence the procedure name) for each of the variables listed in the VAR statement. Most of these statistics are not very useful in the data checking operation. The number of nonmissing observations (N), the number of observations not equal to zero (Num $^\wedge= 0$), and the number of observations greater than zero (Num > 0) are probably the only items that are of interest to you at this time.

One of the most important sections of the PROC UNIVARIATE output, for data checking purposes, is the section labeled "Extremes." Here you see the five highest and five lowest values for each of your variables. For example, for the variable HR (heart rate), there are three possible data errors under the column label "Lowest" (10, 22, and 22) and three possible data errors under the column label "Highest" (208, 210, and 900). Obviously, having knowledge of reasonable values for each of your variables is essential if this information is to be of any use. Next to the listing of the highest and lowest values is the observation number containing this value. What would be more useful would be the patient or subject number you assigned to each patient. This is easily accomplished by adding an ID statement to PROC UNIVARIATE. You list the name of your identifying variable following the keyword ID. The values of this ID variable are then used in addition to the OBS column. (**Note**: In versions of SAS software prior to Version 7, the ID column replaced the OBS column; in versions after Version 7, both the ID column and the OBS column are displayed.)

If you are running Version 7 or later of SAS software, you can include an ODS (output delivery system) statement to limit the PROC UNIVARIATE output to just the table of extreme values. Here are the PROC UNIVARIATE statements with an ID statement added, as well as the ODS statement to limit the output to the five highest and five lowest data values (the "Extremes").

Program 2-4   Adding an ID Statement to PROC UNIVARIATE

```
/*************************************************************\
| The ODS statement is valid for V7 and later.              |
| Note that the name EXTREMEOBS may change in future SAS releases. |
| Use ODS TRACE ON; before the PROC and ODS TRACE OFF; after |
| the PROC to obtain a list of output object names (found in |
| the SAS Log).                                             |
\*************************************************************/

ODS SELECT EXTREMEOBS;

PROC UNIVARIATE DATA=CLEAN.PATIENTS;
   TITLE "Using PROC UNIVARIATE to Look for Outliers";
   ID PATNO;
   VAR HR SBP DBP;
RUN;
```

The section of output showing the "Extremes" for the variable heart rate (HR) follows:

```
Using PROC UNIVARIATE to Look for Outliers


The UNIVARIATE Procedure
Variable:  HR  (Heart Rate)

                  Extreme Observations

    --------Lowest--------          --------Highest-------

  Value    PATNO      Obs        Value    PATNO      Obs

    10     020         23          90                 8
    22     023         25         101     004         4
    22     014         15         208     017        19
    48     022         24         210     008         9
    58     019         20         900     321        22


  Missing Value         .
  Count                 3
  % Count/Nobs       10.00
```

Before the addition of the ID statement, we only had columns labeled VALUE and OBS. With the ID statement there is a new column labeled PATNO that contains the values of your ID variable (PATNO), making it easier to locate the original patient data and check for errors. (**Note**: The column, which contains the values of the ID variable, that here is labeled PATNO will be labeled ID in releases of SAS software prior to Version 7).

The middle section of the output on page 28 contains a stem-and-leaf plot and a box plot. These two data visualizations come from an area of statistics known as exploratory data analysis (EDA). (For an excellent reference see *Exploratory Data Analysis*, by John Tukey, Reading, Massachusetts: Addison-Wesley.) Let's focus on the plots for the variable DBP (diastolic blood pressure). The stem-and-leaf plot can be thought of as a sideways histogram. For this variable, the diastolic blood pressures are grouped in 20-point intervals. For example, the stem labeled "8" represents the values from 80 to 99. Instead of simply placing X's or some other symbol to represent the bar in this sideways histogram, the next digit in the value is used instead. Thus, you see that there were three values of 80, one 82, two 84's, one 88, and two 90's. You can ignore these values and just think of the stem-and-leaf plot as a histogram, or you might be interested in the additional information that the leaf values give you. A quick examination of this plot shows that there were some abnormally low and high diastolic blood pressure values. This useful information complements the "Extremes" information. The "Extremes" only lists the five highest and five lowest values; the stem-and-leaf plot shows all the values in your data set.

To the right of the stem-and-leaf plot is a box plot. This plot shows the mean (+ sign), the median (the dashed line between the two asterisks), and the 25th and 75th percentiles (the bottom and top of the box, respectively). The distance between the top and bottom of the box is called the interquartile range and can also be found earlier in the outout labeled as "Q1-Q3." Extending from both the top and bottom of the box are whiskers and outliers. The whiskers represent data values within one-and-a-half interquartile ranges above or below the box. (**Note**: The EDA people call the top and bottom of the box, the hinges.) Any data values more than one-and-a-half but less than three interquartile ranges above or below the box (hinges) are represented by 0's. Two data values for DBP (8 and 20) fit this description in the box plot on page 28. Finally, any data values more than three interquartile ranges above or below the top and bottom hinges are represented by asterisks. For your DBP variable, two data points, 180 and 200, fit this description. The final graph, called a normal probability plot, is of interest to statisticians and helps determine deviations from a theoretical distribution called a normal or Gaussian distribution. The information displayed in the normal probability plot may not be useful for your data cleaning task because you are looking for data errors and are not particularly interested if the data are normally distributed or not.

Using PROC PRINT with a WHERE Statement to List Invalid Data Values

While PROC MEANS and PROC UNIVARIATE can be useful as a first step in data cleaning for numeric variables, they can produce large volumes of output and may not give you all the information you want, and certainly not in a concise form. One way to check each numeric variable for invalid values is to use PROC PRINT, followed by the appropriate WHERE statement.

Suppose you want to check all the data for any patient having a heart rate outside the range of 40 to 100, a systolic blood pressure outside the range of 80 to 200, and a diastolic blood pressure outside the range of 60 to 120. For this example, missing values are not treated as invalid. The PROC PRINT step in Program 2-5 reports all patients with out-of-range values for heart rate, systolic blood pressure, or diastolic blood pressure.

Program 2-5   Using a WHERE Statement with PROC PRINT to List Out-of-Range Data

```
PROC PRINT DATA=CLEAN.PATIENTS;
   WHERE (HR NOT BETWEEN 40 AND 100 AND HR IS NOT MISSING)       OR
         (SBP NOT BETWEEN 80 AND 200 AND SBP IS NOT MISSING)     OR
         (DBP NOT BETWEEN 60 AND 120 AND DBP IS NOT MISSING);
   TITLE "Out-of-Range Values for Numeric Variables";
   ID PATNO;
   VAR HR SBP DBP;
RUN;
```

You don't need the parentheses in the WHERE statements because the AND operator is evaluated before the OR operator. However, because this author can never seem to remember the order of operation of Boolean operators, the parentheses were included for clarity. Extra parentheses do no harm.

The resulting output is shown next.

```
Out-of-Range Values for Numeric Variables

PATNO     HR    SBP    DBP

 004     101    200    120
 008     210     .      .
 009      86    240    180
 010       .     40    120
 011      68    300     20
 014      22    130     90
 017     208     .      84
 321     900    400    200
 020      10     20      8
 023      22     34     78
```

A disadvantage of this listing is that an observation is printed if one or more of the variables is outside the specified range. To obtain a more precise listing that shows only the data values outside the normal range, you can use a DATA step as described in the next section.

Using a DATA Step to Check for Invalid Values

A simple DATA _NULL_ step can also be used to produce a report on out-of-range values. The approach here is the same as the one described in the section of Chapter 1 that begins on page 6.

Program 2-6   Using a DATA _NULL_ Step to List Out-of-Range Data Values

```
DATA _NULL_;
   INFILE "C:\CLEANING\PATIENTS.TXT" PAD;
   FILE PRINT; ***Send output to the Output window;
   TITLE "Listing of Patient Numbers and Invalid Data Values";
   ***Note: We will only input those variables of interest;
   INPUT @1  PATNO    $3.
         @15 HR       3.
         @18 SBP      3.
         @21 DBP      3.;
   ***Check HR;
   IF (HR LT 40 AND HR NE .) OR HR GT 100 THEN PUT PATNO= HR=;
   ***Check SBP;
   IF (SBP LT 80 AND SBP NE .) OR SBP GT 200 THEN PUT PATNO= SBP=;
   ***Check DBP;
   IF (DBP LT 60 AND DBP NE .) OR DBP GT 120 THEN PUT PATNO= DBP=;
RUN;
```

Here is the output from Program 2-6.

```
Listing of Patient Numbers and Invalid Data Values

PATNO=004   HR=101
PATNO=008   HR=210
PATNO=009   SBP=240
PATNO=009   DBP=180
PATNO=010   SBP=40
PATNO=011   SBP=300
PATNO=011   DBP=20
PATNO=014   HR=22
PATNO=017   HR=208
PATNO=321   HR=900
PATNO=321   SBP=400
PATNO=321   DBP=200
PATNO=020   HR=10
PATNO=020   SBP=20
PATNO=020   DBP=8
PATNO=023   HR=22
PATNO=023   SBP=34
```

Notice that a statement such as "IF HR LT 40" includes missing values because missing values are interpreted by SAS programs as the smallest possible value. Therefore, the following statement

```
IF HR LT 40 OR HR GT 100 THEN PUT PATNO= HR=;
```

will produce a listing that includes missing heart rates as well as out-of-range values (which may be what you want).

Creating a Macro for Range Checking

Because range checking is such a common data cleaning task, it makes some sense to automate the procedure somewhat. Looking at Program 2-6, you see that the range checking lines all look similar except for the variable names and the low and high cutoff values. Even if you are not a macro expert, the following program should not be too difficult to understand. (For an excellent review of macro programming, I recommend two books in SAS Institute's BBU series: *SAS® Macro Programming Made Easy*, by Michele M. Burlew, and *Carpenter's Complete Guide to the SAS® Macro Language*, by Art Carpenter, both published by SAS Institute, Cary, NC.)

Program 2-7   Writing a Macro to List Out-of-Range Data Values

```
*----------------------------------------------------------------*
| Program Name: RANGE.SAS  in C:\CLEANING                         |
| Purpose: Macro that takes lower and upper limits for a         |
|          numeric variable and an ID variable to print out      |
|          an exception report to the Output window.             |
| Arguments: DSN    - Data set name                              |
|            VAR    - Numeric variable to test                   |
|            LOW    - Lowest valid value                         |
|            HIGH   - Highest valid value                        |
|            IDVAR  - ID variable to print in the exception      |
|                       report                                   |
| Example: %RANGE(CLEAN.PATIENTS,HR,40,100,PATNO)                |
*----------------------------------------------------------------*;

%MACRO RANGE(DSN,VAR,LOW,HIGH,IDVAR);

   TITLE "Listing of Invalid Patient Numbers and Data Values";
   DATA _NULL_;
      SET &DSN(KEEP=&IDVAR &VAR);  ❶
      FILE PRINT;
      IF (&VAR LT &LOW AND &VAR NE .) OR &VAR GT &HIGH THEN
         PUT "&IDVAR:" &IDVAR  @18 "Variable:&VAR"
                               @38 "Value:" &VAR
                               @50 "out-of-range";
   RUN;

%MEND RANGE;
```

First, a brief explanation. A macro program is a piece of SAS code where parts of the code are substituted with variable information by the macro processor before the code is processed in the usual way by the SAS compiler. The macro in Program 2-7 is named RANGE, and it begins with a %MACRO statement and ends with a %MEND (macro end) statement. The first line of the macro contains the macro name, followed by a list of arguments. When the macro is called, the macro processor replaces each of these arguments with the values you specify. Then, in the macro program, every macro variable (that is, every variable name preceded by an ampersand (&)) is replaced by the assigned value. For example, if you want to use this macro to look for out-of-range values for heart rate in the PATIENTS data set, you would call the macro like this

```
%RANGE(CLEAN.PATIENTS,HR,40,100,PATNO)
```

The macro processor will substitute these calling arguments for the &variables in the macro program. For example, line ❶ will become:

```
SET CLEAN.PATIENTS(KEEP=PATNO HR);
```

&DSN was replaced by CLEAN.PATIENTS, &IDVAR was replaced by PATNO, and &VAR was replaced by HR. To be sure this concept is clear, (and to help you understand how the macro processor works), you can call the macro with the MPRINT option turned on. This option lists the macro generated code in the SAS Log. Here is the section of the SAS Log containing the macro generated statements when the macro is called with the above arguments:

```
MPRINT(RANGE):  TITLE "Listing of Invalid Patient Numbers and Data
Values";
MPRINT(RANGE):  DATA _NULL_;
MPRINT(RANGE):  SET CLEAN.PATIENTS(KEEP=PATNO HR);
MPRINT(RANGE):  FILE PRINT;
MPRINT(RANGE):  IF (HR LT 40 AND HR NE .) OR HR GT 100 THEN PUT "PATNO:"
                PATNO @18 "Variable:HR" @38 "Value:" HR @50 "out-of-
range";
MPRINT(RANGE):  RUN;
```

By the way, the missing semicolon at the end of the line where the macro is called is not a mistake — you don't need it. The reason is that the macro code contains a semicolon after the last RUN statement so that an extra semicolon is unnecessary. If you like, you may put one in any way. As pointed out by Mike Zdeb, one of my reviewers, if you include the unnecessary semicolon, you can change the line to a comment by placing an asterisk at the beginning.

The results from running the macro with the above calling arguments are listed next:

```
Listing of Invalid Patient Numbers and Data Values

PATNO:004        Variable:HR        Value:101    out-of-range
PATNO:008        Variable:HR        Value:210    out-of-range
PATNO:014        Variable:HR        Value:22     out-of-range
PATNO:017        Variable:HR        Value:208    out-of-range
PATNO:321        Variable:HR        Value:900    out-of-range
PATNO:020        Variable:HR        Value:10     out-of-range
PATNO:023        Variable:HR        Value:22     out-of-range
```

While this saves on programming time, it is not as efficient as a program that checks all the numeric variables in one DATA step.   However, sometimes it is reasonable to sacrifice computer time for human time.

Using Formats to Check for Invalid Values

Just as you did with character values in Chapter 1, you can use user-defined formats to check for out-of-range data values. Program 2-8 uses formats to find invalid data values, based on the same ranges used in Program 2-5 in this chapter.

Program 2-8   Detecting Out-of-Range Values Using User-Defined Formats

```
PROC FORMAT;
   VALUE HR_CK   40-100,  . = 'OK';
   VALUE SBP_CK 80-200,  . = 'OK';
   VALUE DBP_CK 60-120,  . = 'OK';
RUN;

DATA _NULL_;
   INFILE "C:\CLEANING\PATIENTS.TXT" PAD;
   FILE PRINT; ***Send output to the Output window;
   TITLE "Listing of Invalid Patient Numbers and Data Values";
   ***Note: We will only input those variables of interest;
   INPUT @1  PATNO    $3.
          @15 HR        3.
          @18 SBP       3.
          @21 DBP       3.;
   IF PUT(HR,HR_CK.)   NE 'OK' THEN PUT PATNO= HR=;
   IF PUT(SBP,SBP_CK.) NE 'OK' THEN PUT PATNO= SBP=;
   IF PUT(DBP,DBP_CK.) NE 'OK' THEN PUT PATNO= DBP=;
RUN;
```

This is a fairly simple and efficient program. The user-defined formats HR_CK., SBP_CK., and DBP_CK. all assign the formatted value 'OK' for any data value in the acceptable range. In the DATA step, the result of the PUT function is the value of the first argument (the variable to be tested) formatted by the format specified as the second calling argument of the function. For example, any value of heart rate between 40 and 100 (or missing) falls into the format range 'OK'. A value of 22 for heart rate does not fall within the range of 40 to 100 or missing and the formatted value 'OK' is not assigned. In that case, the PUT function for heart rate does not return the value 'OK' and the IF statement condition is true. The appropriate PUT statement is then executed and the invalid value is printed to the print file.

Output from this program is shown next:

```
Listing of Invalid Patient Numbers and Data Values

PATNO=004   HR=101
PATNO=008   HR=210
PATNO=009   SBP=240
PATNO=009   DBP=180
PATNO=010   SBP=40
PATNO=011   SBP=300
PATNO=011   DBP=20
PATNO=014   HR=22
PATNO=017   HR=208
PATNO=321   HR=900
PATNO=321   SBP=400
PATNO=321   DBP=200
PATNO=020   HR=10
PATNO=020   SBP=20
PATNO=020   DBP=8
PATNO=023   HR=22
PATNO=023   SBP=34
```

Notice that patient number 27, who had a value of 'NA' for heart rate, did not appear in this listing. Why not? Well, the INPUT statement generates a missing value in its attempt to read a character value with a numeric informat. Because missing values are not treated as errors in this example, no error listing is produced for patient number 27. If you would like to include invalid character values (such as NA) as errors, you can use the internal _ERROR_ variable to check if such a value was processed by the INPUT statement. Unfortunately, the program cannot tell which variable for patient number 27 contained the invalid value. It is certainly possible to distinguish between invalid character values in numeric fields from true missing values. One possible approach is to use an enhanced numeric informat. Another is to read all of the numeric variables as character data, test the values, and then convert to numeric for range checking. In the section that follows, the program demonstrates how a user-defined enhanced numeric informat can be used. A simple "work-around" for program 2-8 is to test for any character values that were converted to missing values by using the internal variable _ERROR_, which gets set to '1' any time the input processor detects such an error. A modified version of Program 2-8, shown below, will print a notification that one or more variables for a patient had an invalid character value.

Program 2-9   Modifying the Previous Program to Detect Invalid (Character) Data Values

```
DATA _NULL_;
   INFILE "C:\CLEANING\PATIENTS.TXT" PAD;
   FILE PRINT; ***Send output to the Output window;
   TITLE "Listing of Invalid Patient Numbers and Data Values";
   ***Note: We will only input those variables of interest;
   INPUT @1  PATNO    $3.
         @15 HR       3.
         @18 SBP      3.
         @21 DBP      3.;
   IF PUT(HR,HR_CK.)   NE 'OK' OR _ERROR_ GT 0 THEN PUT PATNO= HR=;
   IF PUT(SBP,SBP_CK.) NE 'OK' OR _ERROR_ GT 0 THEN PUT PATNO= SBP=;
   IF PUT(DBP,DBP_CK.) NE 'OK' OR _ERROR_ GT 0 THEN PUT PATNO= DBP=;
   IF _ERROR_ GT 0 THEN
      PUT PATNO= "had one or more invalid character values";
   ***Set the Error flag back to 0;
   _ERROR_ = 0;
RUN;
```

Using Informats to Check for Invalid Values

You can accomplish the same result as the previous program by using user-defined informats. Remember that informats are used to replace values as the raw data is being read in or as the second argument in an INPUT function. Following is a program very similar to Program 2-9, however this one uses informats.

Program 2-10 Using User-Defined Informats to Detect Out-of-Range Data Values

```
PROC FORMAT;
    INVALUE HR_CK  40-100, . = 9999;
    INVALUE SBP_CK 80-200, . = 9999;
    INVALUE DBP_CK 60-120, . = 9999;
RUN;

DATA _NULL_;
    INFILE "C:\CLEANING\PATIENTS.TXT" PAD;
    FILE PRINT; ***Send output to the Output window;
    TITLE "Listing of Invalid Patient Numbers and Data Values";
    ***Note: We will only input those variables of interest;
    INPUT @1  PATNO    $3.
          @15 HR       HR_CK3.
          @18 SBP      SBP_CK3.
          @21 DBP      DBP_CK3.;
    IF HR NE 9999 THEN PUT PATNO= HR=;
    IF SBP NE 9999 THEN PUT PATNO= SBP=;
    IF DBP NE 9999 THEN PUT PATNO= DBP=;
RUN;
```

PROC FORMAT is used to create three informats (note the use of INVALUE statements instead of the usual VALUE statements). For the informat HR_CK, any numeric value in the range 40 to 100 or missing is assigned a value of 9999. Note that you cannot assign a character value here because the result of a numeric informat must be numeric. In this example, using the value of 9999 is a good choice because 9999 can never be a valid value for any of the variables (they are stored in three columns in the input file).

Running Program 2-10 results in the following output:

```
Listing of Invalid Patient Numbers and Data Values

PATNO=004   HR=101
PATNO=008   HR=210
PATNO=009   SBP=240
PATNO=009   DBP=180
PATNO=010   SBP=40
PATNO=011   SBP=300
PATNO=011   DBP=20
PATNO=014   HR=22
PATNO=017   HR=208
PATNO=321   HR=900
PATNO=321   SBP=400
PATNO=321   DBP=200
PATNO=020   HR=10
PATNO=020   SBP=20
PATNO=020   DBP=8
PATNO=023   HR=22
PATNO=023   SBP=34
PATNO=027   HR=.
```

If you look carefully at the output from this program and the earlier program that used user-defined formats, you will notice that patient number 027 is listed here with a missing heart rate but not shown in the earlier listing. What's going on? Inspection of the raw data shows a value of 'NA' for heart rate for patient number 27. When you used a format, the original data value of 'NA' was converted to a numeric missing value by the SAS processor (with a resulting message being written to the Log). The result of the PUT function was therefore 'OK' and the value was not flagged as invalid. When an informat was used, the value 'NA' was not in the valid range so that the value of 9999 was not assigned to heart rate and the value was flagged as invalid. If you would like to go the "extra mile," you can use an enhanced numeric informat to assign a value to any alphabetic value. With this technique, you can distinguish invalid character values from true missing values. Program 2-11 demonstrates this.

Program 2-11 Modifying the Previous Program to Detect Invalid (Character) Data Values

```
PROC FORMAT;
   INVALUE HR_CK (UPCASE)
                   40 - 100, .  = 9999
                   'A' - 'Z'    = 8888;
   INVALUE SBP_CK (UPCASE)
                   80 - 200, .  = 9999
                   'A' - 'Z'    = 8888;
   INVALUE DBP_CK (UPCASE)
                   60 - 120, .  = 9999
                   'A' - 'Z'    = 8888;
RUN;


DATA _NULL_;
   INFILE "C:\CLEANING\PATIENTS.TXT" PAD;
   FILE PRINT; ***Send output to the Output window;
   TITLE "Listing of Invalid Patient Numbers and Data Values";
   ***Note: We will only input those variables of interest;
   INPUT @1  PATNO     $3.
         @15 HR        HR_CK3.
         @18 SBP       SBP_CK3.
         @21 DBP       DBP_CK3.;
   IF HR = 8888 THEN PUT PATNO= "Invalid character value for HR";
   ELSE IF HR NE 9999 THEN PUT PATNO= HR=;

   IF SBP = 8888 THEN PUT PATNO= "Invalid character value for SBP";
   ELSE IF SBP NE 9999 THEN PUT PATNO= SBP=;

   IF DBP = 8888 THEN PUT PATNO= "Invalid character value for DBP";
   ELSE IF DBP NE 9999 THEN PUT PATNO= DBP=;
RUN;
```

The UPCASE option converts any character values to uppercase before it is determined if the value fits into one of the specified ranges. Notice that the ranges for the three informats contain both numeric ranges and character ranges. This feature, called an enhanced numeric informat, is very powerful and allows programs to read a combination of numeric and character data with a single informat (see SAS Technical Report P-222, *Changes and Enhancements to Base SAS Software, Release 6.07*). Notice in the next output, that patient number 27 is reported to have invalid character data for heart rate.

```
Listing of Invalid Patient Numbers and Data Values

PATNO=004 HR=101
PATNO=008 HR=210
PATNO=009 SBP=240
PATNO=009 DBP=180
PATNO=010 SBP=40
PATNO=011 SBP=300
PATNO=011 DBP=20
PATNO=014 HR=22
PATNO=017 HR=208
PATNO=321 HR=900
PATNO=321 SBP=400
PATNO=321 DBP=200
PATNO=020 HR=10
PATNO=020 SBP=20
PATNO=020 DBP=8
PATNO=023 HR=22
PATNO=023 SBP=34
PATNO=027 Invalid character value for HR
```

Using PROC UNIVARIATE to Look for Highest and Lowest Values by Percentage

Let's return to the problem of locating the "n" highest and "n" lowest values for each of several numeric variables in the data set. Remember that earlier in this chapter, you used PROC UNIVARIATE to list the five highest and five lowest values for your three numeric variables. First of all, this procedure prints lots of other statistics that you don't need (or want), unless you use the output delivery system to limit the output.

If you are running a version of SAS software prior to Version 7 or you want to control the number of high and low values to list, you can write a custom program to give you exactly what you want. The approach is to have PROC UNIVARIATE output a data set containing the cutoff values on the lower and upper range of interest. The first program described lists the bottom and top "n" percent of the values. Next, the program is turned into a macro so that it is easier to use.

Program 2-12 uses PROC UNIVARIATE to print out the bottom and top "n" percent of the data values.

Program 2-12 Using PROC UNIVARIATE to Print the Top and Bottom "n" Percent of Data Values

```
***Solution using PROC UNIVARIATE and Percentiles;

LIBNAME CLEAN "C:\CLEANING";
***The two macro variables that follow define the lower and upper
   percentile cut points;

***Change the value in the line below to the percentile cut-off
   you want;
%LET LOW_PER=20;   ❶

***Compute the upper cut-off value;
%LET UP_PER= %EVAL(100 - &LOW_PER);   ❷

***Choose a variable to operate on;
%LET VAR = HR;   ❸

PROC UNIVARIATE DATA=CLEAN.PATIENTS NOPRINT;   ❹
   VAR &VAR;
   ID PATNO;
   OUTPUT OUT=TMP PCTLPTS=&LOW_PER &UP_PER PCTLPRE = L_;   ❺
RUN;

DATA HILO;
   SET CLEAN.PATIENTS(KEEP=PATNO &VAR);   ❻
   ***Bring in upper and lower cutoffs for variable;
   IF _N_ = 1 THEN SET TMP;   ❼
   IF &VAR LE L_&LOW_PER THEN DO;
      RANGE = 'LOW ';
      OUTPUT;
   END;
   ELSE IF &VAR GE L_&UP_PER THEN DO;
      RANGE = 'HIGH';
      OUTPUT;
   END;
RUN;

PROC SORT DATA=HILO(WHERE=(&VAR NE .));   ❽
   BY DESCENDING RANGE &VAR;
RUN;
```

```
PROC PRINT DATA=HILO;
   TITLE "High and Low Values for Variables";
   ID PATNO;
   VAR RANGE &VAR;
RUN;
```

Let's go through this program step by step. To make the program somewhat general, it uses several macro variables. Line ❶ assigns the lower percentile to a macro variable (LOW_PER) using a %PUT statement. Line ❷ computes the upper percentile cutoff (UP_PER) by subtracting the lower percentile cutoff from 100. (Note: The %EVAL function is needed here to perform the integer arithmetic. If the value of LOW_PER was 20, the value of &UP_PER, without the %EVAL function, would be the text string "100 − 20" instead of 80.) If you look at line ❺, you see the two macro variables LOW_PER and UP_PER preceded by an ampersand (&). As discussed earlier, before the SAS processor runs any SAS program, it runs the macro processor, which processes all the macro statements and substitutes the assigned values of the macro variables. In this program, after the macro processor does its job, line ❺ reads:

```
OUTPUT OUT=TMP PCTLPTS=20 80 PCTLPRE = L_;
```

That is, the two macro variables, &LOW_PER and &UP_PER are replaced by the values assigned by the %LET statements, 20 and 80 respectively.

In line ❸, a macro variable (VAR) is assigned the value of one of the numeric variables to be checked (HR). To run this program on another numeric variable, SBP for example, you only have to change the variable name in line ❸.

PROC UNIVARIATE can be used to create an output data set containing information that is normally printed out by the procedure. Because you only want the output data set and not the listing from the procedure, use the NOPRINT option as shown in line ❹. As you did before, you are supplying PROC UNIVARIATE with an ID statement so that the ID variable (PATNO in this case) will be included in the output data set. Line ❺ defines the name of the output data set and specifies the information you want it to include. The keyword OUT= names your data set (TMP) and PCTLPTS= instructs the program to create two variables; one to hold the value of the VAR variable at the 20th percentile and the other for the 80th percentile. In order for this procedure to create the variable names for these two variables, the keyword PCTLPRE= (percentile prefix) is used. Because you set the prefix to L_, the procedure creates two variables, L_20 and L_80.

The cut points you choose are combined with your choice of prefix to create these two variable names. The data set TMP contains only one observation and three variables, PATNO (because of the ID statement), L_20, and L_80. The value of L_20 is 58 and the value of _80 is 88, the 20th and 80th percentile cutoffs, respectively. The remainder of the program is easier to follow.

You want to add the two values of L_20 and L_80 to every observation in the original PATIENTS data set. Let's do this with a "trick." The SET statement in line ❻ brings in an observation from the PATIENTS data set, keeping only the variables PATNO and HR (because the macro variable &VAR was set to HR). Line ❼ is executed only on the first iteration of this DATA step (when _N_ is equal to 1). Because all variables brought in with a SET statement are automatically retained, the values for L_20 and L_80 are added to every observation in the data set HILO.

Finally, for each observation coming in from the PATIENTS data set, the value of HR is compared to the lower and upper cutoff points defined by L_20 and L_80. If the value of HR is at or below the value of L_20, RANGE is set to the value 'LOW' and the observation is added to the data set HILO. Likewise, if the value of HR is at or above the value of L_80, RANGE is set to 'HIGH' and the observation is added to the data set HILO. Before you print out the contents of the data set HILO, you sort it first ❽ so that the low values and high values are grouped, and within these groups, the values sorted from lowest to highest. The keyword DESCENDING is used in the first level sort so that the LOW values are listed before the HIGH values ('H' comes before 'L' in a normal ascending alphabetical sort). Within each of these two groups, the data values are listed from low to high. It would probably be nicer for the HIGH values to be listed from highest to lowest, but it would not be worth the effort. The final listing from this program is shown next.

```
High and Low Values for Variables

PATNO     RANGE      HR

 020      LOW         10
 014      LOW         22
 023      LOW         22
 022      LOW         48
 003      LOW         58
 019      LOW         58
 001      HIGH        88
 007      HIGH        88
          HIGH        90
 004      HIGH       101
 017      HIGH       208
 008      HIGH       210
 321      HIGH       900
```

To turn the above program into a macro is actually quite straightforward. The macro version is shown in Program 2-13.

Program 2-13 Creating a Macro to List the Highest and Lowest "n" Percent of the Data Using PROC UNIVARIATE

```
*----------------------------------------------------------------*
| Program Name: HILOWPER.SAS  in C:\CLEANING                     |
| Purpose: To list the n percent highest and lowest values for  |
|          a selected variable.                                  |
| Arguments: DSN     - Data set name                             |
|            VAR     - Numeric variable to test                  |
|            PERCENT - Upper and Lower percentile cutoff         |
|            IDVAR   - ID variable to print in the report        |
| Example: %HILOWPER(CLEAN.PATIENTS,SBP,20,PATNO)                |
*----------------------------------------------------------------*;

%MACRO HILOWPER(DSN,VAR,PERCENT,IDVAR);

   ***Compute upper percentile cutoff;
   %LET UP_PER = %EVAL(100 - &PERCENT);

   PROC UNIVARIATE DATA=&DSN NOPRINT;
      VAR &VAR;
      ID &IDVAR;
      OUTPUT OUT=TMP PCTLPTS=&PERCENT &UP_PER PCTLPRE = L_;
   RUN;

   DATA HILO;
      SET &DSN(KEEP=&IDVAR &VAR);
      IF _N_ = 1 THEN SET TMP;
      IF &VAR LE L_&PERCENT THEN DO;
         RANGE = 'LOW ';
         OUTPUT;
      END;
      ELSE IF &VAR GE L_&UP_PER THEN DO;
         RANGE = 'HIGH';
         OUTPUT;
      END;
   RUN;

   PROC SORT DATA=HILO(WHERE=(&VAR NE .));
      BY DESCENDING RANGE &VAR;
   RUN;
```

```
PROC PRINT DATA=HILO;
   TITLE "Low and High Values for Variables";
   ID &IDVAR;
   VAR RANGE &VAR;
RUN;

PROC DATASETS LIBRARY=WORK NOLIST;
   DELETE TMP;
   DELETE HILO;
RUN;
QUIT;
```

```
%MEND HILOWPER ;
```

The only change, besides the four macro variables, is the addition of PROC DATASETS to delete the two temporary data sets TMP and HILO. To demonstrate this macro, the three lines below call the macro to list the highest and lowest 20 % of the values for heart rate (HR), systolic blood pressure (SBP), and diastolic blood pressure (DBP) in the data set PATIENTS.

```
%HILOWPER(CLEAN.PATIENTS,HR,20,PATNO)
%HILOWPER(CLEAN.PATIENTS,SBP,20,PATNO)
%HILOWPER(CLEAN.PATIENTS,DBP,20,PATNO)
```

Using PROC RANK to Look for Highest and Lowest Values by Percentage

There is a simpler and more efficient way to list the highest and lowest "n" percent of the data values, that is, by using PROC RANK. The reason that the previous, more complicated program was shown, is that it produces a slightly more accurate listing than the program shown in this section. PROC RANK is designed to produce a new variable (or replace the values of an existing variable) with values equal to the ranks of another variable. For example, if the variable X has values of 7, 3, 2, and 8, the equivalent ranks would be 3, 2, 1, and 4, respectively. However, PROC RANK has a very useful option (GROUPS=) that allows you to group your data values. For example, if you set GROUPS=4, the new variable that usually holds the rank values, will now have values of 0, 1, 2, and 3, with those observations in groups 0 being in the bottom quartile and observations in group 3 being in the top quartile. So, if you want to print out the top 20% of your data values, you set the GROUPS option to 5, each group representing 20% of your data values. The bottom 20% corresponds to the ranked variable having a value of 0, and the top 20% corresponds to the ranked variable having a value of 4. (Yes, it is odd that without the GROUPS= option ranks go from 1 to n and with the GROUPS= option, the groups go from 0 to n − 1.)

Now let's see how to apply this idea to a program that will list the top and bottom "n" percent of your data values. Because you have already seen a program and macro that lists highest and lowest "n" percent of your data values, only the macro version is shown here.

Program 2-14 Creating a Macro to List the Highest and Lowest "n" Percent of the Data Using PROC RANK

```
*-------------------------------------------------------------------*
| Macro Name: HI_LOW_P                                              |
| Purpose: To list the upper and lower n% of values                 |
| Arguments: DSN      - Data set name (one- or two-level            |
|            VAR      - Variable to test                            |
|            PERCENT - Upper and lower n%                            |
|            IDVAR    - ID variable                                 |
| Example: %HI_LOW_P(CLEAN.PATIENTS,SBP,20,PATNO)                   |
*-------------------------------------------------------------------*;

%MACRO HI_LOW_P(DSN,VAR,PERCENT,IDVAR);
   ***Compute number of groups for PROC RANK;
   %LET GRP = %SYSEVALF(100 / &PERCENT,FLOOR);   ❶
   ***Value of the highest GROUP from PROC RANK, equal to the
      number of groups - 1;
   %LET TOP = %EVAL(&GRP - 1);   ❷

   PROC FORMAT;   ❸
      VALUE RNK 0='Low' &TOP='High';
   RUN;

   PROC RANK DATA=&DSN OUT=NEW GROUPS=&GRP;   ❹
      VAR &VAR;
      RANKS RANGE;
   RUN;

   ***Sort and keep top and bottom n%;
   PROC SORT DATA=NEW (WHERE=(RANGE IN (0,&TOP)));   ❺
      BY &VAR;
   RUN;
```

```
***Produce the report;
PROC PRINT DATA=NEW;    ❻
TITLE "Upper and Lower &PERCENT.% Values for %UPCASE(&VAR)";
   ID &IDVAR;
   VAR RANGE &VAR;
   FORMAT RANGE RNK.;
RUN;

PROC DATASETS LIBRARY=WORK NOLIST;    ❼
   DELETE NEW;
RUN;
QUIT;


%MEND HI_LOW_P;
```

First, you need to compute the approximate number of groups that will correspond to the percentage you want. Line ❶ uses the %SYSEVALF function to do this computation. This function, unlike its companion %EVAL, allows noninteger arithmetic and also provides various conversions (CEIL, FLOOR, INTEGER, or BOOLEAN) for the results. The floor conversion was chosen because you would rather have the program list too many values (i.e., a smaller value for the GROUPS= option) than too few. For example, if you want the top and bottom 8% of your data values, the value of GRP would be FLOOR(100/8) = 12 and the value for TOP would be 11. It is this rounding that may produce a slightly less accurate report than the program that uses PROC UNIVARIATE. The RNK format assigns the formats 'Low' and 'High' to the ranked variable.

The key to the whole program is PROC RANK ❹, which uses the GROUPS= option to divide the data values into groups. The sort ❺ accomplishes two things: 1) It subsets the data set with the WHERE data set option, keeping only the top and bottom groups, and 2) it puts the data values in order from the smallest to the largest. All that is left to do is to print the report ❻ and delete the temporary data set ❼.

Issue the following statement to see a list of the top and bottom 10% of your values for SBP (systolic blood pressure):

```
%HI_LOW_P(CLEAN.PATIENTS,SBP,10,PATNO)
```

This produces the following output:

```
Upper and Lower 10% Values for SBP

PATNO     RANGE     SBP

  020     Low        20
  023     Low        34
  011     High      300
  321     High      400
```

Extending PROC RANK to Look for Highest and Lowest "n" Values

Instead of listing the highest and lowest "n" percent of the data values, you might want to select the cutoffs based on the actual number of values, not the percent. This is slightly harder because you have to determine the number of observations in the data set and to compute the percentage cutoffs, given the number of values you want. To save some time (and space) only the macro version of this program is presented. It is followed by the explanation.

Program 2-15 Creating a Macro to List the Top and Bottom "n" Data Values Using PROC RANK

```
*----------------------------------------------------------------*
| Macro Name: HI_LOW_N                                            |
| Purpose: To list N highest and lowest values (approximately)    |
| Arguments: DSN      - Data set name (one- or two-level          |
|            VAR      - Variable to test                          |
|            N        - Number of highest and lowest values       |
|            IDVAR    - ID variable                               |
| Example: %HI_LOW_N (CLEAN.PATIENTS,SBP,10,PATNO)                |
*----------------------------------------------------------------*;
```

```
%MACRO HI_LOW_N(DSN,VAR,N,IDVAR);
   ***Find the number of observations in data set;
   %LET DSID = %SYSFUNC(OPEN(&DSN));    ❶
   %LET N_OBS = %SYSFUNC(ATTRN(&DSID,NOBS));
   %LET RETURN = %SYSFUNC(CLOSE(&DSID));

   ***Compute number of groups, from N and N_OBS;
   %LET GRP = %SYSEVALF(&N_OBS / &N,FLOOR);   ❷
   ***Continue as in the macro based on percents;
   %LET TOP = %EVAL(&GRP - 1);

   PROC FORMAT;
      VALUE RNK 0='Low' &TOP='High';
   RUN;


   PROC RANK DATA=&DSN OUT=NEW GROUPS=&GRP;
      VAR &VAR;
      RANKS RANGE;
   RUN;

   ***Sort and keep top and bottom n%;
   PROC SORT DATA=NEW (WHERE=(RANGE IN (0,&TOP)));
      BY  &VAR;
   RUN;

   ***Produce the report;
   PROC PRINT DATA=NEW;
     TITLE "Approximate Highest and Lowest &N Values for %UPCASE(&VAR)";
      ID &IDVAR;
      VAR RANGE &VAR;
      FORMAT RANGE RNK.;
   RUN;

   PROC DATASETS LIBRARY=WORK NOLIST;
      DELETE NEW;
   RUN;
   QUIT;

%MEND HI_LOW_N;
```

This macro is very similar to the macro in the previous section, except that the number of groups computed is based on the number of observations in the data set. If there are too many missing values for the variable of interest, the number of nonmissing values for that variable can be used instead of the number of observations in the entire data set. This number can be determined by using PROC MEANS to output the number of nonmissing values to a data set.

This macro uses %SYSFUNC ❶ to open the data set and determine the number of observations, and close the data set (don't forget to close data sets!). (%SYSFUNC executes SAS language functions and returns the results to the macro facility.) The calculation of the number of groups is accomplished in line ❷. The FLOOR function is used so that we err on the side of too few groups (more data values listed) rather than too many. The remainder of the program is identical to Program 2-14.

To demonstrate this macro, let's list the 10 highest and 10 lowest systolic blood pressure readings in the PATIENTS data set by using the following statement:

```
%HI_LOW_N (CLEAN.PATIENTS,SBP,10,PATNO)
```

The resulting output is shown next.

```
Approximate Highest and Lowest 10 Values for SBP

PATNO     RANGE     SBP

 020      Low        20
 023      Low        34
 010      Low        40
 006      Low       102
 025      Low       102
 013      Low       108
 003      Low       112
 022      Low       114
 019      Low       118
 028      High      150
 027      High      166
 003      High      190
          High      190
 004      High      200
 009      High      240
 011      High      300
 321      High      400
```

Notice that there are nine low values and eight high values. This discrepancy from the number we selected will be less in larger data sets, but it still may not be exact because of rounding errors and the way that groups are selected when the number of data values is not an exact multiple of the number of groups. Also, PROC RANK assigns ranks only to nonmissing values. Program 2-17 and Program 2-18 both provide listings of exactly "n" highest and lowest values, but, at the expense of processing time.

If you have large numbers of missing values for variables in your data set, there is an alternative and more CPU intensive method to determine the number of nonmissing values instead of the number of observations in the data set, as shown in Program 2-16. Just substitute the following lines for the three %SYSFUNC lines in the previous macro.

Program 2-16 Determining the Number of Nonmissing Observations in a Data Set

```
***Find the number of nonmissing observations in data set;
PROC MEANS DATA=&DSN NOPRINT;
   VAR &VAR;
   OUTPUT OUT=TMP N=NONMISS;
RUN;

DATA _NULL_;
   SET TMP;
   ***Assign the value of NONMISS to the macro variable N_OBS;
   CALL SYMPUT("N_OBS",NONMISS);
RUN;
```

If you use this code, add the data set TMP to the list of data sets to delete at the end of the macro. If you use this method of determining the number of nonmissing observations in your data set, the same macro call that was used in Program 2-15 produces output with 13 low values and 14 high values. (The computation of the percentiles gives a larger value if the denominator is the number of observations with nonmissing values rather than the total number of observations.)

Finding Another Way to Determine Highest and Lowest Values

There is usually more than one way to solve any SAS problem. Here is another approach to listing the 10 highest and 10 lowest values for a variable in a SAS data set. The advantage of this program is that it always gives you exactly 10 high and 10 low values. The program is presented first, followed by a macro version of it.

Program 2-17 Listing the Highest and Lowest "n" Values Using PROC SORT

```
LIBNAME CLEAN "C:\CLEANING";
%LET VAR = HR;  ***Assign values to two macro variables;
%LET IDVAR = PATNO;

PROC SORT DATA=CLEAN.PATIENTS(KEEP=&IDVAR &VAR      ❶
                              WHERE=(&VAR NE .))
                              OUT=TMP;
   BY &VAR;
RUN;

DATA _NULL_;
   TITLE "Ten Highest and Ten Lowest Values for &VAR";
   SET TMP NOBS=NUM_OBS;      ❷
   HIGH = NUM_OBS - 9;      ❸
   FILE PRINT;

   IF _N_ LE 10 THEN DO;      ❹
     IF _N_ = 1 THEN PUT / "Ten Lowest Values" ;
     PUT "&IDVAR = " &IDVAR @15 &VAR;
   END;

   IF _N_ GE HIGH THEN DO;      ❺
     IF _N_ = HIGH THEN PUT / "Ten Highest Values" ;
     PUT "&IDVAR = " &IDVAR @15 &VAR;
   END;
RUN;
```

This is a simpler program than the one that uses PROC RANK. One drawback, however, is that the data set needs to be sorted each time the program is run for a different variable. This may be OK for a relatively small data set but inappropriate (and inefficient) for a large one. In this program, only the variable name and the ID variable are assigned to macro variables. To make the program as efficient as possible, a KEEP= data set option is used with PROC SORT ❶. In addition, only the nonmissing observations are placed in the sorted temporary data set TMP (because of the WHERE= data set option). The data set TMP will contain only the ID variable and the variable to be checked, in order, from lowest to highest. Therefore, the first 10 observations in this data set are the 10 lowest, nonmissing values for the variable to be checked. Use the NOBS= option in the SET statement (line ❷) to obtain the number of observations in the data set TMP. Because this data set only contains nonmissing values, the 10 highest values for your variable start with observation NUM_OBS - 9. This program uses a DATA _NULL_ and PUT statements to provide the listing of high and low values. As an alternative, you could create a temporary data set and use PROC PRINT to provide the listing.

One final note: this program does not check if there are fewer than 20 nonmissing observations for the variable to be checked. That would probably be overkill. If you had that few observations, you wouldn't really need a program at all, just a PROC PRINT!

Running Program 2-17 on the PATIENTS data set for the heart rate variable (HR) produces the following:

```
Ten Highest and Ten Lowest Values for HR

Ten Lowest Values
PATNO = 020    10
PATNO = 014    22
PATNO = 023    22
PATNO = 022    48
PATNO = 003    58
PATNO = 019    58
PATNO = 012    60
PATNO = 123    60
PATNO = 028    66
PATNO = 003    68

Ten Highest Values
PATNO = 002    84
PATNO = 002    84
PATNO = 009    86
PATNO = 001    88
PATNO = 007    88
PATNO =        90
PATNO = 004    101
PATNO = 017    208
PATNO = 008    210
PATNO = 321    900
```

A macro version of this program is straightforward (see Program 2-18). To make it more general, the data set name is added as a macro variable as well.

Program 2-18 Creating a Macro to List the "n" Highest and Lowest Data Values Using PROC SORT

```
*-----------------------------------------------------------------*
| Program Name: TEN.SAS  in C:\CLEANING                           |
| Purpose: To list the 10 highest and lowest data values for     |
|          a variable in a SAS data set using DATA step processing |
| Arguments: DSN    - Data set name                              |
|            VAR    - Numeric variable to be checked             |
|            IDVAR  - ID variable name                           |
|                                                                 |
| Example: %TEN(CLEAN.PATIENTS,HR,PATNO)                         |
*-----------------------------------------------------------------*;

%MACRO TEN(DSN,VAR,IDVAR);

   PROC SORT DATA=&DSN(KEEP=&IDVAR &VAR
                           WHERE=(&VAR NE .))
                      OUT=TMP;
      BY &VAR;
   RUN;

   DATA _NULL_;
      TITLE "Ten Highest and Ten Lowest Values for %UPCASE(&VAR)";
      SET TMP NOBS=NUM_OBS;
      HIGH = NUM_OBS - 9;
      FILE PRINT;

      IF _N_ LE 10 THEN DO;
         IF _N_ = 1 THEN PUT / "Ten Lowest Values" ;
         PUT "&IDVAR = " &IDVAR @15 "&VAR = " &VAR;
      END;

      IF _N_ GE HIGH THEN DO;
         IF _N_ = HIGH THEN PUT / "Ten Highest Values" ;
         PUT "&IDVAR = " &IDVAR @15 "&VAR = " &VAR;
      END;

   RUN;

%MEND TEN;
```

Checking a Range Using an Algorithm Based on Standard Deviation

One way of deciding what constitutes reasonable cutoffs for low and high data values is to use an algorithm based on the data values themselves. For example, you could decide to flag all values more than two standard deviations from the mean. However, if you had some severe data errors, the standard deviation could be so badly inflated that obviously incorrect data values might lie within two standard deviations. A possible workaround for this would be to compute the standard deviation after removing some of the highest and lowest values. For example, you could compute a standard deviation of the middle 50% of your data and use this to decide on outliers. Another popular alternative is to use an algorithm based on the interquartile range (the difference between the 25th percentile and the 75th percentile). Some programs and macros based on these ideas are presented in the next two sections.

Let's first see how you could identify data values more than two standard deviations from the mean. You can use PROC MEANS to compute the standard deviations and a short DATA step to select the outliers, as shown in Program 2-19.

Program 2-19   Detecting Outliers Based on the Standard Deviation

```
LIBNAME CLEAN "C:\CLEANING";
***Output means and standard deviations to a data set;
PROC MEANS DATA=CLEAN.PATIENTS NOPRINT;
   VAR HR SBP DBP;
   OUTPUT OUT=MEANS(DROP=_TYPE_ _FREQ_)
          MEAN=M_HR M_SBP M_DBP
          STD=S_HR S_SBP S_DBP;
RUN;
```

```
%LET N_SD = 2;
*** The number of standard deviations to list;


DATA _NULL_;
   FILE PRINT;
   TITLE "Statistics for Numeric Variables";
   SET CLEAN.PATIENTS;
   ***Bring in the means and standard deviations;
   IF _N_ = 1 THEN SET MEANS;
   ARRAY RAW[3] HR SBP DBP;
   ARRAY _MEAN[3] M_HR M_SBP M_DBP;
   ARRAY _STD[3] S_HR S_SBP S_DBP;

   DO I = 1 TO DIM(RAW);
      IF RAW[I] LT _MEAN[I] - &N_SD*_STD[I] AND RAW[I] NE .
      OR RAW[I] GT _MEAN[I] + &N_SD*_STD[I] THEN PUT PATNO= RAW[I]=;
   END;
RUN;
```

The PROC MEANS step computes the mean and standard deviation for each of the numeric variables in your data set. To make the program more flexible, the number of standard deviations above or below the mean that you would like to report is assigned to a macro variable (N_SD). To compare each of the raw data values against the limits defined by the mean and standard deviation, you need to combine the values in the single observation data set created by PROC MEANS to the original data set. You use the same trick you used earlier, that is, you execute a SET statement only once, when _N_ is equal to one. Because all the variables brought into the program data vector (PDV) with a SET statement are retained, these summary values will be available in each observation in the PATIENTS data set. Finally, to save some typing, three arrays were created to hold the original raw variables, the means, and the standard deviations, respectively. The IF statement at the bottom of this DATA step prints out the ID variable and the raw data value for any value above or below the designated cutoff.

The results of running this program on the PATIENTS data set with N_SD set to two follows:

```
Statistics for Numeric Variables

PATNO=009 DBP=180
PATNO=011 SBP=300
PATNO=321 HR=900
PATNO=321 SBP=400
PATNO=321 DBP=200
PATNO=020 DBP=8
```

How would you go about computing cutoffs based on the middle 50% of your data? Calculating a mean and standard deviation on the middle 50% of the data (called trimmed statistics by robust statisticians — and I know some statisticians that are very robust!) is easy if you first use PROC RANK (with a GROUPS= option) to identify quartiles, and then use this information in a subsequent PROC MEANS step to compute the mean and standard deviation of the middle 50% of your data. Your decision on how many of these trimmed standard deviation units should be used to define outliers is somewhat of a trial-and-error process. Obviously, (well, maybe not that obvious) the standard deviation computed on the middle 50% of your data will be smaller than the standard deviation computed from all of your data if you have outliers. The difference between the two will be even larger if there are some dramatic outliers in your data. (This will be demonstrated later in this section.) As an approximation, if your data are normally distributed, the trimmed standard deviation is approximately 2.6 times smaller than the untrimmed value. So, if your original cutoff was plus or minus two standard deviations, you might choose 5 or 5.2 trimmed standard deviations as your cutoff scores. What follows is a program that computes trimmed statistics and uses them to identify outliers.

Program 2-20 Detecting Outliers Based on a Trimmed Mean

```
PROC RANK DATA=CLEAN.PATIENTS OUT=TMP GROUPS=4;
   VAR HR;
   RANKS R_HR;
RUN;

PROC MEANS DATA=TMP NOPRINT;
   WHERE R_HR IN (1,2);   ***The middle 50%;
   VAR HR;
   OUTPUT OUT=MEANS(DROP=_TYPE_ _FREQ_)
          MEAN=M_HR
          STD=S_HR;
RUN;
```

```
DATA _NULL_;
   TITLE "Outliers Based on Trimmed Standard Deviation";
   FILE PRINT;

   %LET N_SD = 5.25;
   ***The value of 5.25 computed from the trimmed mean is
      approximately equivalent to the 2 standard deviations
      you used before, computed from all the data. Set this
      value approximately 2.65 times larger than the number
      of standard deviations you would compute from untrimmed data;

   SET CLEAN.PATIENTS;
   IF _N_ = 1 THEN SET MEANS;
   IF HR LT M_HR - &N_SD*S_HR AND HR NE .
      OR HR GT M_HR + &N_SD*S_HR THEN PUT PATNO= HR=;
RUN;
```

There is one slight complication here, compared to the earlier nontrimmed version of the program. The middle 50% of the observations can be different for each of the numeric variables you want to test. So, if you want to run the program for several variables, it would be convenient to assign to a macro variable the name of the numeric variable that will be tested. This is done next, but first, a brief explanation of the program. PROC RANK is used with the GROUPS= option to create a new variable (R_HR), which will have values of 0, 1, 2, or 3, depending on which quartile the value lies. Because you want both the original value for HR and the rank value, use a RANKS statement, which allows you to give a new name to the variable that will hold the rank of the variable listed in the VAR statement. All that is left to do is to run PROC MEANS as you did before, except that a WHERE statement selects the middle 50% of the data values. What follows is the same as Program 2-19, except that arrays are not needed because you can only process one variable at a time. Finally, here is the output from Program 2-20.

```
Outliers Based on Trimmed Standard Deviation

PATNO=008   HR=210
PATNO=014   HR=22
PATNO=017   HR=208
PATNO=321   HR=900
PATNO=020   HR=10
PATNO=023   HR=22
```

Notice that the method based on a nontrimmed standard deviation reported only one HR as an outlier (PATNO=321, HR=900) while the method based on a trimmed mean identified six values. The reason? The heart rate value of 900 inflated the nontrimmed standard deviation so much that none of the other values fell within two standard deviations.

Macros Based on the Two Methods of Outlier Detection

It is straightforward to turn each of the above programs into more general purpose macros. First, here is a macro that detects outliers based on a standard deviation computed from all the data.

Program 2-21 Creating a Macro to Detect Outliers Based on a Standard Deviation

```
*-------------------------------------------------------------------*
| Program Name: SD_ALL.SAS   in C:\CLEANING                          |
| Purpose: To identify outliers based on n standard deviations       |
|       from the mean.                                               |
| Arguments: DSN    - Data set name                                  |
|            VAR    - Numeric variable to be checked                 |
|            IDVAR  - ID variable name                               |
|            N_SD   - The number of standard deviation units for     |
|                     declaring an outlier                           |
|                                                                    |
| Example: %SD_ALL(CLEAN.PATIENTS,HR,PATNO,2)                        |
*-------------------------------------------------------------------*;
%MACRO SD_ALL(DSN,VAR,IDVAR,N_SD);

   TITLE1 "Outliers for Variable &VAR Data Set &DSN";
   TITLE2 "Based on &N_SD Standard Deviations";

   PROC MEANS DATA=&DSN NOPRINT;
   VAR &VAR ;
   OUTPUT OUT=MEANS(DROP=_TYPE_ _FREQ_)
          MEAN=M
          STD=S;
   RUN;

   DATA _NULL_;
      FILE PRINT;
      SET &DSN;
      IF _N_ = 1 THEN SET MEANS;
         IF &VAR LT M - &N_SD*S AND &VAR NE .
         OR &VAR GT M + &N_SD*S THEN PUT &IDVAR= &VAR=;
   RUN;
```

```
PROC DATASETS LIBRARY=WORK NOLIST;
   DELETE MEANS;
RUN;
QUIT;


%MEND SD_ALL;
```

Next, a macro is shown that detects outliers based on computing the mean and standard deviation from the middle 50% of the data values. It can be easily modified to use more or less data by adjusting the GROUPS= option in PROC RANK and modifying the WHERE statement in the PROC MEANS step as appropriate.

Program 2-22 Creating a Macro to Detect Outliers Based on a Trimmed Mean

```
*-----------------------------------------------------------------*
| Program Name: SD_TRIM.SAS  in C:\CLEANING                       |
| Purpose: To identify outliers based on n standard deviations    |
|          from the mean, computed from the middle 50% of the data. |
| Arguments: DSN    - Data set name                               |
|            VAR    - Numeric variable to be checked              |
|            IDVAR  - ID variable name                            |
|            N_SD   - The number of standard deviation units you  |
|                     would specify if the data values were not   |
|                     trimmed.                                    |
|                                                                 |
| EXAMPLE: %SD_TRIM(CLEAN.PATIENTS,HR,PATNO,2)                    |
*-----------------------------------------------------------------*;
%MACRO SD_TRIM(DSN,VAR,IDVAR,N_SD);

   TITLE1 "Outliers for Variable &VAR Data Set &DSN";
   TITLE2 "Based on &N_SD Standard Deviations Estimated from Trimmed
          (50%)Data";

   PROC RANK DATA=&DSN OUT=TMP GROUPS=4;
      VAR &VAR;
      RANKS R;
   RUN;

   PROC MEANS DATA=TMP NOPRINT;
      WHERE R IN (1,2);  ***The middle 50%;
      VAR &VAR;
      OUTPUT OUT=MEANS(DROP=_TYPE_ _FREQ_)
             MEAN=M
             STD=S;
   RUN;
```

```
DATA _NULL_;
   FILE PRINT;
   SET &DSN;
   IF _N_ = 1 THEN SET MEANS;
   IF &VAR LT M - &N_SD*S*2.65 AND &VAR NE .
      OR &VAR GT M + &N_SD*S*2.65 THEN PUT &IDVAR= &VAR=;   ❶
RUN;

PROC DATASETS LIBRARY=WORK NOLIST;
   DELETE MEANS;
   DELETE TMP;
RUN;
QUIT;


%MEND SD_TRIM;
```

Notice in line ❶ of the above macro, the value 2.65 is the estimated amount you need to inflate the trimmed standard deviation to estimate the untrimmed standard deviation for normally distributed data.

Demonstrating the Difference between the Two Methods

To show the difference between these two methods of outlier detection, look at the following small data set:

```
DATA TRIM;
   INPUT X @@;
   PATNO + 1;
DATALINES;
1.02 1.06 1.23 2.00 1.09 1.15 1.23 1.33 1.99 1.11
1.45 156  4.88 2.11 1.54 1.64 1.73 1.19 1.21 1.29
;
```

First, a brief comment on the program. Ordinarily, SAS goes to a new line for each INPUT statement in a DATA step and for each iteration of the DATA step. The double at sign (@@) prevents this from happening and allows you to place data for multiple observations on a single line. This is a convenient way to save some space. Next, remember that the SAS statement

```
PATNO + 1;
```

does several things. First, it initializes PATNO at 0. Second, PATNO is automatically retained, and third, each time the statement executes, PATNO is incremented by 1.

There are 20 values of X in the data set TRIM. One of the values, 156, is a data error and should have been 1.56. This type of error is not all that uncommon. The other suspicious value is 4.88. This may or may not be a data error. You would probably want your data checking program to flag the 4.88 value so that it could be checked.

Using the following two lines to run both macros,

```
%SD_ALL(TRIM,PATNO,X,2)
```

```
%SD_TRIM(TRIM,PATNO,X,2)
```

the resulting output is

```
Outliers for Variable X Data Set TRIM
Based on Two Standard Deviations

PATNO=12 X=156

Outliers for Variable X Data Set TRIM
Based on Two Standard Deviations Estimated from Trimmed (50%) Data

PATNO=12 X=156
PATNO=13 X=4.88
```

The program based on the standard deviation of all the values, with a cutoff of two standard deviations only lists the value of 156. Why didn't the program identify 4.88 as an outlier? Because of the 156, the mean and standard deviation, using all 20 of the data values, were 9.31 and 34.54, respectively. The single value being approximately 100 times larger than the other values grossly inflated the mean and standard deviation. The value of 4.88 is less than one standard deviation from the sample mean. The mean and standard deviation of the trimmed data set are 1.38 and .195, respectively. Using these values, the value of 4.88 is easily identified as a possible outlier.

Checking a Range Based on the Interquartile Range

Yet another way to look for outliers is a method devised by advocates of exploratory data analysis (EDA). This is a robust method, much like the previous method described, based on a trimmed mean. It uses the interquartile range (the distance from the 25th percentile to the 75th percentile) and defines an outlier as a multiple of the interquartile range above or below the upper or lower hinge, respectively. For those not familiar with EDA terminology, the lower hinge is the value corresponding to the 25th percentile (the value below which 25% of the data values lie).   The upper hinge is the value corresponding to the 75% percentile. For example, you may want to examine any data values more than two interquartile ranges above the upper hinge or below the lower hinge. This is an attractive method because it is independent of the distribution of the data values.

An easy way to determine the interquartile range and the upper and lower hinges is to use PROC UNIVARIATE to output these quantities. Presented next is a macro, which is similar to the one in the previous section, but this one uses the number of interquartile ranges instead of an estimate of the standard deviation.

Program 2-23 Detecting Outliers Based on the Interquartile Range

```
*-------------------------------------------------------------------*
| Program Name: INTER_Q.SAS   in C:\CLEANING                        |
| Purpose: To identify outliers based on n interquartile ranges     |
| Arguments: DSN    - Data set name                                 |
|            VAR    - Numeric variable to be checked                |
|            IDVAR  - ID variable name                              |
|            N_IQR  - The number of interquartile ranges above or   |
|                     below the upper and lower hinge (75th and     |
|                     25th percentile points) to declare a value    |
|                     an outlier.                                   |
|                                                                   |
| Example: %INTER_Q(CLEAN.PATIENTS,HR,PATNO,2)                      |
*-------------------------------------------------------------------*;
%MACRO INTER_Q(DSN,VAR,IDVAR,N_IQR);
   PROC UNIVARIATE DATA=&DSN NOPRINT;
      VAR &VAR;
      OUTPUT OUT=TMP Q3=UPPER Q1=LOWER QRANGE=IQR;   ❶
   RUN;

   DATA _NULL_;
      TITLE "Outliers Based on &N_IQR Interquartile Ranges";
      FILE PRINT;
      SET &DSN;
      IF _N_ = 1 THEN SET TMP;
      IF &VAR LT LOWER - &N_IQR*IQR AND &VAR NE .
         OR &VAR GT UPPER + &N_IQR*IQR THEN PUT &IDVAR= &VAR=;   ❷
   RUN;
```

```
    PROC DATASETS LIBRARY=WORK NOLIST;
        DELETE TMP;
    RUN;
    QUIT;

%MEND INTER_Q;
```

Use PROC UNIVARIATE to output the values of the 25th and 75th percentile to a data set ❶. In the DATA _NULL_ step that follows, any values more than "n" interquartile ranges (the macro variable N_IQR) below the lower hinge or above the upper hinge are flagged as errors and reported ❷.

To demonstrate this macro, the calling sequence below checks for outliers more than two interquartile ranges above or below the upper or lower hinge, respectively. The calling statement is

```
%INTER_Q(CLEAN.PATIENTS,HR,PATNO,2)
```

with the resulting output shown next.

```
Outliers Based on Two Interquartile Ranges

PATNO=008    HR=210
PATNO=017    HR=208
PATNO=321    HR=900
```

The same macro, used on the data set TRIM (on page 64) with the number of interquartile ranges set at two, results in the output shown next.

```
Outliers Based on Two Interquartile Ranges

PATNO=12 X=156
PATNO=13 X=4.88
```

Notice that both the values 4.88 and 156 are identified in this method.

Checking Ranges for Several Variables

In this final section of this chapter, the range checking macro developed on page 35 is expanded to do two things: One, make the macro more flexible so that it can either treat missing values as valid or invalid; and two, allow the macro to be called multiple times with different numeric variables and produce one consolidated report when finished. The macro is listed first, followed by a step-by-step explanation.

Program 2-24 Writing a Program to Summarize Data Errors on Several Variables

```
*-----------------------------------------------------------------*
| PROGRAM NAME: ERRORSN.SAS  IN C:\CLEANING                        |
| PURPOSE: Accumulates errors for numeric variables in a SAS       |
|          data set for later reporting.                           |
|          This macro can be called several times with a           |
|          different variable each time. The resulting errors      |
|          are accumulated in a temporary SAS data set called      |
|          ERRORS.                                                 |
| ARGUMENTS: DSN     - SAS data set name (assigned with a %LET)    |
|            IDVAR   - ID variable (assigned with a %LET)          |
|                                                                  |
|            VAR     - The variable name to test                   |
|            LOW     - Lowest valid value                          |
|            HIGH    - Highest valid value                         |
|            M       - Missing value flag.  If=1 count missing     |
|                      values as invalid, =0, missing values OK    |
|                                                                  |
| EXAMPLE: %LET DSN = CLEAN.PATIENTS;                              |
|          %LET IDVAR = PATNO;                                     |
|          %ERRORSN(HR,40,100,1)                                   |
|          %ERRORSN(SBP,80,200,0)                                  |
|          %ERRORSN(DBP,60,120,0)                                  |
|          Test the numeric variables HR, SBP, and DBP in the      |
|          data set CLEAN.PATIENTS for data outside the ranges     |
|          40 to 100, 80 to 200, and 60 to 120, respectively.      |
|          The ID variable is PATNO and missing values are to      |
|          be flagged as invalid for HR but not for SBP or DBP.    |
*-----------------------------------------------------------------*;
```

```
LIBNAME CLEAN "C:\CLEANING";

%LET DSN=CLEAN.PATIENTS;  ***Define Data set name and;  ❶
%LET IDVAR=PATNO;         ***ID variable;

%MACRO ERRORSN(VAR,LOW,HIGH,M);  ❷

   DATA TMP;
      SET &DSN(KEEP=&IDVAR &VAR);  ❸
      LENGTH REASON $ 10 VARIABLE $ 8;  ❹
      VARIABLE = "&VAR";
      VALUE = &VAR;  ❺
      IF &VAR LT &LOW AND &VAR NE . THEN DO;  ❻
         REASON='LOW';
         OUTPUT;
      END;
      ELSE IF &VAR EQ . AND &M THEN DO;  ❼
         REASON='MISSING';
         OUTPUT;
      END;
      ELSE IF &VAR GT &HIGH THEN DO;  ❽
         REASON='HIGH';
         OUTPUT;
      END;
      DROP &VAR;
   RUN;

   PROC APPEND BASE=ERRORS DATA=TMP;  ❾
   RUN;
   TITLE "Listing Of Errors In Data Set &DATA ";

%MEND ERRORSN;

***Error Reporting Macro - to be run after ERRORSN has been called
   as many times as desired for each numeric variable to be tested;

%MACRO E_REPORT    ❿
   PROC SORT DATA=ERRORS;  ⓫
      BY & IDVAR;
   RUN;
```

```
PROC PRINT DATA=ERRORS;
   TITLE "Error Report for Data Set &DSN";
   ID &IDVAR;
   VAR VARIABLE VALUE REASON;
RUN;

PROC DATASETS LIBRARY=WORK NOLIST; ⓬
   DELETE ERRORS;
   DELETE TMP;
RUN;
QUIT;
```

```
%MEND E_REPORT;
```

To avoid having to enter the data set name and the ID variable each time this macro is called, the two macro variables DSN and IDVAR are assigned with %LET statements ❶. Calling arguments to the macro ❷ are the name of the numeric variable to be tested, the lower and upper valid values for this variable, and a variable to determine if missing values are to be listed in the error report or not. To keep the macro somewhat efficient, only the variable in question and the ID variable are added to the TMP data set because of the KEEP= data set option in line ❸. The variables REASON and VARIABLE ❹ hold values for why the observation was selected and the name of the variable being tested. Because the name of the numeric variable to be tested changes each time the macro is called, a variable called VALUE ❺ is assigned the value of the numeric variable. The range checking is accomplished in lines ❻ and ❽. Line ❼ reports missing values as invalid if the macro variable M is set to 1, otherwise missing values are not treated as errors. Finally, each error found is added to the temporary data set ERRORS by using PROC APPEND ❾. This is the most efficient method of adding observations to an existing SAS data set. Each time the ERRORSN macro is called, all the invalid observations will be added to the ERRORS data set.

The second macro, E_REPORT ❿, is a macro that should be called once after the ERRORSN macro has been called for each of the desired numeric variable range checks. The E_REPORT macro is simple. It sorts the ERRORS data set by the ID variable, so that all errors for a particular ID will be grouped together⓫. Finally, as you have done in the past, use PROC DATASETS ⓬ to clean up the WORK data sets that were created.

To demonstrate how these two macros work, the ERRORSN macro is called three times, for the variables heart rate (HR), systolic blood pressure (SBP), and diastolic blood pressure (DBP), respectively. For the HR variable, you want missing values to appear in the error report; for the other two variables, you do not want missing values listed as errors. Here is the calling sequence:

```
***Calling the ERRORSN macro;
LIBNAME CLEAN.PATIENTS;
%LET DSN = CLEAN.PATIENTS; ***Set two macro variables;
%LET ID = PATNO;

%ERRORSN(HR,40,100,1)
%ERRORSN(SBP,80,200,0)
%ERRORSN(DBP,60,120,0)

***Generate the report;
%E_REPORT
```

And finally, the report that is produced:

```
Error Report for Data Set CLEAN.PATIENTS

PATNO     VARIABLE     VALUE     REASON

 004         HR         101       HIGH
 008         HR         210       HIGH
 009         SBP        240       HIGH
 009         DBP        180       HIGH
 010         HR          .        MISSING
 010         SBP         40       LOW
 011         SBP        300       HIGH
 011         DBP         20       LOW
 014         HR          22       LOW
 017         HR         208       HIGH
 020         HR          10       LOW
 020         SBP         20       LOW
 020         DBP          8       LOW
 023         HR          22       LOW
 023         SBP         34       LOW
 027         HR          .        MISSING
 029         HR          .        MISSING
 321         HR         900       HIGH
 321         SBP        400       HIGH
 321         DBP        200       HIGH
```

The clear advantage of this technique is that it provides a report that lists all the out-of-range or missing value errors for each patient, all in one place.

3   Checking for Missing Values

Introduction

Many data sets contain missing values. There are several ways in which missing values can enter a SAS data set. First of all, the raw data value may be missing, either intentionally or accidentally. Next, an invalid value can cause a missing value to be created. For example, reading a character value with a numeric informat will generate a missing value. Invalid dates are another common cause of SAS generated missing values. Finally, many operations, such as assignment statements, can create missing values. This chapter investigates ways to detect and count missing values for numeric and character variables.

Inspecting the SAS Log

It is vitally important to carefully inspect the SAS Log, especially when creating a SAS data set for the first time. A log filled with messages about invalid data values is a clue that something may be wrong, either with the data or the program. If you know that a numeric field contains invalid character values, you may choose to read those data values with a character informat and to perform a character-to-numeric conversion (using the INPUT function) yourself. This will keep the SAS

Log cleaner and make it easier to spot unexpected errors. Let's look at portions of the SAS Log that were generated when the PATIENTS data set was created.

```
1     LIBNAME CLEAN "C:\CLEANING";
NOTE: Libref CLEAN was successfully assigned as follows:
      Engine:        V7
      Physical Name: C:\CLEANING
2
3     DATA CLEAN.PATIENTS;
4        INFILE "C:\CLEANING\PATIENTS.TXT" PAD;
5        INPUT @1  PATNO    $3.
6              @4  GENDER   $1.
7              @5  VISIT    MMDDYY10.
8              @15 HR       3.
9              @18 SBP      3.
10             @21 DBP      3.
11             @24 DX       $3.
12             @27 AE       $1.;
13
14       LABEL PATNO   = "Patient Number"
15             GENDER  = "Gender"
16             VISIT   = "Visit Date"
17             HR      = "Heart Rate"
18             SBP     = "Systolic Blood Pressure"
19             DBP     = "Diastolic Blood Pressure"
20             DX      = "Diagnosis Code"
21             AE      = "Adverse Event?";
22
23       FORMAT VISIT MMDDYY10.;
24
25    RUN;

NOTE: The infile "C:\CLEANING\PATIENTS.TXT" is:
      File Name=C:\CLEANING\PATIENTS.TXT,
      RECFM=V,LRECL=256

NOTE: Invalid data for VISIT in line 7 5-14.
RULE: ---+---1---+---2---+---3---+---4---+---5---+---6---+---7---+---8
7    007M08/32/1998 88148102   0
RULE: ---+---1---+---2---+---3---+---4---+---5---+---6---+---7---+---8
     92
     183
PATNO=007 GENDER=M VISIT=. HR=88 SBP=148 DBP=102 DX=  AE=0 _ERROR_=1 _N_=7
```

```
NOTE: Invalid data for VISIT in line 12 5-14.
12        011M13/13/1998 68300 20   41
      92
     183
PATNO=011 GENDER=M VISIT=. HR=68 SBP=300 DBP=20 DX=4 AE=1 _ERROR_=1 _N_=12
NOTE: Invalid data for VISIT in line 21 5-14.
21        123M15/12/1999 60         10
      92
     183
PATNO=123 GENDER=M VISIT=. HR=60 SBP=. DBP=. DX=1 AE=0 _ERROR_=1 _N_=21
NOTE: Invalid data for VISIT in line 23 5-14.
23        020F99/99/9999 10 20  8   0
      92
     183
PATNO=020 GENDER=F VISIT=. HR=10 SBP=20 DBP=8 DX=  AE=0 _ERROR_=1 _N_=23
NOTE: Invalid data for VISIT in line 28 5-14.
NOTE: Invalid data for HR in line 28 15-17.
28        027FNOTAVAIL  NA 166106  70
      92
     183
PATNO=027 GENDER=F VISIT=. HR=. SBP=166 DBP=106 DX=7 AE=0 _ERROR_=1 _N_=28
NOTE: 31 records were read from the infile "C:\CLEANING\PATIENTS.TXT".
      The minimum record length was 26.
      The maximum record length was 27.
NOTE: The data set CLEAN.PATIENTS has 31 observations and 8 variables.
NOTE: DATA statement used:
      real time            0.50 seconds
```

The first invalid data message is generated by an invalid date (08/32/1998). This will be discussed in more detail in Chapter 4, "Working with Dates". For now, realize that a numeric missing value (remember that dates are stored as numeric values) will be generated as a result of this invalid date. Several more invalid date messages follow. A missing value for heart rate (HR) was generated for patient number 20 because of the character value 'NA' (not available or not applicable) that was entered. Before going any further, the invalid dates need to be checked, and a decision needs to be made concerning the 'NA' value for heart rate.

Using PROC MEANS and PROC FREQ to Count Missing Values

There are several procedures that will count missing values for you. It may be normal to have missing values for certain variables in your data set. There may also be variables for which no missing values are permitted (such as a patient ID). An easy way to count missing values is by using PROC MEANS; for character variables, PROC FREQ will provide this information. Program 3-1 is a simple program that can be used to check the number of numeric and character missing values in the PATIENTS data set.

Program 3-1   Counting Missing and Nonmissing Values for Numeric and Character Variables

```
LIBNAME CLEAN "C:\CLEANING";

TITLE "Missing Value Check for the PATIENTS Data Set";

PROC MEANS DATA=CLEAN.PATIENTS N NMISS;
RUN;

PROC FORMAT;
   VALUE $MISSCNT ' '   = 'MISSING'
                OTHER = 'NONMISSING';
RUN;

PROC FREQ DATA=CLEAN.PATIENTS;
   TABLES _CHARACTER_ / NOCUM MISSING;
   FORMAT _CHARACTER_ $MISSCNT.;
RUN;
```

The check for numeric missing values is straightforward. By using the N and NMISS options with PROC MEANS, you get a count of the nonmissing and missing values for all your numeric variables (the default if no VAR statement is included). You could also choose to use a VAR statement to list only the variables of interest.

Counting missing values for character variables takes an extra step. First, you do not simply want to create one-way frequencies for all the character variables. Some variables, such as patient ID (PATNO) can, conceivably, have thousands of values. By creating a character format that has only two value ranges, one for missing and the other for everything else, you can have PROC FREQ count missing and nonmissing values for you.

Notice also, that it is necessary to use the SAS keyword _CHARACTER_ in the TABLES statement (or to provide a list of character variables). PROC FREQ can produce frequency tables for numeric as well as character variables. Finally, the TABLES option MISSING includes the missing values in the body of the frequency listing. Examination of the listing from these two procedures is a good first step in your investigation of missing values. The output from Program 3-1 is shown next.

```
Missing Value Check for the PATIENTS Data Set

The MEANS Procedure

                                                        N
Variable    Label                             N      Miss
----------------------------------------------------------
VISIT       Visit Date                       24         7
HR          Heart Rate                       28         3
SBP         Systolic Blood Pressure          27         4
DBP         Diastolic Blood Pressure         28         3
----------------------------------------------------------

Missing Value Check for the PATIENTS data set

The FREQ Procedure

              Patient Number

PATNO           Frequency        Percent
-----------------------------------------
MISSING             1             3.23
NONMISSING         30            96.77


                 Gender

GENDER          Frequency        Percent
-----------------------------------------
MISSING             1             3.23
NONMISSING         30            96.77


              Diagnosis Code

DX              Frequency        Percent
-----------------------------------------
MISSING             8            25.81
NONMISSING         23            74.19


              Adverse Event?

AE              Frequency        Percent
-----------------------------------------
MISSING             1             3.23
NONMISSING         30            96.77
```

Using DATA Step Approaches to Identify and Count Missing Values

Counting missing values is not usually enough. If you have variables for which missing values are not allowed, you need to locate the observations so that the original data values can be checked and the errors corrected. A simple DATA step with a PUT statement is one approach. Program 3-2 checks for any missing visit dates, heart rates (HR), or adverse events (AE).

Program 3-2   Writing a Simple DATA Step to List Missing Data Values and an ID Variable

```
DATA _NULL_;
   INFILE "C:\CLEANING\PATIENTS.TXT" PAD;
   FILE PRINT; ***Send output to the Output window;
   TITLE "Listing of Missing Values";
   ***Note: We will only input those variables of interest;
    INPUT @1  PATNO    $3.
       @5  VISIT     MMDDYY10.
       @15 HR        3.
       @27 AE        $1.;
   IF VISIT = .  THEN PUT "Missing or invalid visit date for ID " PATNO;
   IF HR    = .  THEN PUT "Missing or invalid HR for ID " PATNO;
   IF AE    = ' ' THEN PUT "Missing or invalid AE for ID " PATNO;
RUN;
```

Output from running Program 3-2 is shown next.

```
Listing of Missing Values

Missing or invalid visit date for ID 007
Missing or invalid HR for ID 010
Missing or invalid visit date for ID 011
Missing or invalid AE for ID 013
Missing or invalid visit date for ID 015
Missing or invalid visit date for ID 123
Missing or invalid visit date for ID 321
Missing or invalid visit date for ID 020
Missing or invalid visit date for ID 027
Missing or invalid HR for ID 027
Missing or invalid HR for ID 029
```

What do you do about missing patient numbers? Obviously, you can't list which patient number is missing because you don't have that information. One possibility is to report the patient number or numbers preceding the missing number (in the original order of data entry). If you sort the data set first, all the missing values will "float" to the top and you will not have a clue as to which patients they belong to. Here is a program that prints out the two previous patient ID's when a missing ID is found.

Program 3-3 Attempting to Locate a Missing or Invalid Patient ID by Listing the Two Previous ID's

```
DATA _NULL_;
   SET CLEAN.PATIENTS;
   ***Be sure to run this on the unsorted data set;
   FILE PRINT;
   TITLE "Listing of Missing Patient Numbers";
   PREV_ID = LAG(PATNO);
   PREV2_ID = LAG2(PATNO);
   IF PATNO = ' ' THEN PUT "Missing Patient ID. Two previous ID's are:"
      PREV2_ID "and " PREV_ID / @5 "Missing Record is number " _N_;
   ELSE IF INPUT(PATNO,?? 3.)  = . THEN
      PUT "Invalid Patient ID:" PATNO +(-1)". Two previous ID's are:"
      PREV2_ID "and " PREV_ID / @5 "Missing Record is number " _N_;
RUN;
```

Although there are several solutions to listing the patient numbers from the preceding observations, the LAG function serves the purpose here. Remember to execute the LAG and LAG2 functions for every observation. Then, when a missing patient number is encountered, the two lagged variables will be the ID's from the previous two observations. The assumption in this program is that there are not more than three missing patient numbers in a row. If that is a possibility, you could list more than two previous patient ID's or include patient ID's following the missing one as well. Notice that we added the observation number to the output by printing the value of the internal SAS variable _N_. This provides one additional clue in finding the missing patient number.

Here is the output from Program 3-3:

```
Listing of Missing Patient Numbers

Invalid Patient ID:XX5. Two previous ID's are:003 and 004
    Missing Record is number 5
Missing Patient ID. Two previous ID's are:006 and 007
    Missing Record is number 8
```

Another approach is to list the values of all the variables for any missing or invalid patient ID. This may give a clue to the identity of the missing ID. Using PROC PRINT with a WHERE statement makes this an easy task, as demonstrated by the SAS code in Program 3-4.

Program 3-4   Using PROC PRINT to List Data for Missing or Invalid Patient ID's

```
PROC PRINT DATA=CLEAN.PATIENTS;
   TITLE "Data Listing for Patients with Missing or Invalid ID's";
   WHERE PATNO = ' ' OR INPUT(PATNO,3.) = .;
RUN;
```

Here is the corresponding output.

```
Data Listing for Patients with Missing or Invalid ID's

Obs    PATNO    GENDER         VISIT    HR    SBP    DBP    DX    AE

 5      XX5       M        05/07/1998   68    120     80     1     0
 8                M        11/11/1998   90    190    100           0
```

Before leaving this section on DATA step detection of missing values, let's modify Program 3-2, which listed missing dates, heart rates, and adverse events, to count the number of each missing variable as well.

Program 3-5 Listing and Counting Missing Values for Selected Variables

```
DATA _NULL_;
   INFILE "C:\CLEANING\PATIENTS.TXT" PAD END=LAST;
   FILE PRINT; ***Send output to the Output window;
   TITLE "Listing of Missing Values";
   ***Note: We will only input those variables of interest;
   INPUT @1  PATNO      $3.
       @5  VISIT      MMDDYY10.
       @15 HR         3.
       @27 AE         $1.;
   IF VISIT = . THEN DO;
      PUT "Missing or invalid visit date for ID " PATNO;
      N_VISIT + 1;
   END;
   IF HR    = . THEN DO;
      PUT "Missing or invalid HR for ID " PATNO;
      N_HR + 1;
   END;
   IF AE    = ' ' THEN DO;
      PUT "Missing or invalid AE for ID " PATNO;
      N_AE + 1;
   END;

   IF LAST THEN
         PUT // "Summary of missing values" /
         25*'-' /
         "Number of missing dates = " N_VISIT /
         "Number of missing HR's = " N_HR /
         "Number of missing adverse events = " N_AE;
RUN;
```

Each time a missing value is located, the respective missing counter is incremented by 1. Because you only want to see the totals once after all the data lines have been read, use the END= option in the INFILE statement to create the logical variable LAST. LAST will be true when the last record is being read from the raw data file PATIENTS.TXT. So, in addition to the earlier listing, you have the additional lines of output shown next.

```
Summary of missing values
--------------------------
Number of missing dates = 7
Number of missing HR's = 3
Number of missing adverse events = 1
```

Using **PROC TABULATE** to Count Missing and Nonmissing Values for Numeric Variables

Instead of using PROC MEANS to count numeric missing values, you might want a more attractive table from PROC TABULATE. This, however, adds significant complications. Before writing the general purpose program, let's first see how PROC TABULATE can provide information on your numeric variables. Program 3-6 lists the number of nonmissing and missing values and the minimum and maximum values for three numeric variables in the PATIENTS data set.

Program 3-6   Listing the Number of Nonmissing and Missing Values and the Minimum and Maximum Values for All Numeric Variables

```
PROC TABULATE DATA=CLEAN.PATIENTS FORMAT=8.;
   TITLE "Missing Values, Low and High Values for Numeric Variables";
   VAR HR SBP DBP;
   TABLE HR SBP DBP,
         N NMISS MIN MAX / RTSPACE=26;
   KEYLABEL N       = 'Number'
            NMISS   = 'Number Missing'
            MIN     = 'Lowest Value'
            MAX     = 'Highest Value';
RUN;
```

The option FORMAT=8. provides a default format for all the printed numbers. Eight was chosen because it allows the column headings to fit nicely. The option RTSPACE (row-title-space) leaves enough room for the variable labels. Finally, the KEYLABEL statement lets you choose more convenient names for the generated statistics. The output from Program 3-6 follows. (Remember that the labels for the three numeric variables are stored with the permanent data set.)

Missing Values, Low and High Values for Numeric Variables				
	Number	Number Missing	Lowest Value	Highest Value
Heart Rate	28	3	10	900
Systolic Blood Pressure	27	4	20	400
Diastolic Blood Pressure	28	5	8	200

Using PROC TABULATE to Count Missing and Nonmissing Values for Character Variables

A similar program can provide the number of missing and nonmissing values for the character variables in the data set. There are some differences in how PROC TABULATE can be used to count missing and nonmissing values for character variables. You need to apply a format having only two categories, missing and nonmissing, to each of the character variables to accomplish your goal; the same trick used with PROC FREQ earlier in this chapter. In addition, use the MISSING option to force the number of missing values to be listed.

Program 3-7   Using PROC TABULATE to Count Missing and Nonmissing Values for Character Variables

```
PROC FORMAT;
   VALUE $MISSCH ' ' = 'Missing'
               OTHER = 'Nonmissing';
RUN;

PROC TABULATE DATA=CLEAN.PATIENTS MISSING FORMAT=8.;
   CLASS PATNO DX AE;
   TABLE PATNO DX AE,
       N / RTSPACE=26;
   FORMAT PATNO DX AE $MISSCH.;
   KEYLABEL N    = 'Number';
RUN;
```

The only statistic you can request, when there are no analysis variables listed, is N, the number of observations in each of the formatted categories. Because the character format $MISSCH. has only two levels representing missing and nonmissing values respectively, you "trick" the procedure into giving you what you want. The output from Program 3-7 follows.

Missing Values, Low and High Values for Character Variables	
	Number
Patient Number	
Missing	1
Nonmissing	30
Diagnosis Code	
Missing	8
Nonmissing	23
Adverse Event?	
Missing	1
Nonmissing	30

Creating a General Purpose Macro to Count Missing and Nonmissing Values for Both Numeric and Character Variables

When you omit the VAR statement (or use _NUMERIC_ in place of the variable list) with PROC MEANS, all numeric variables are listed. With PROC FREQ, you can use _CHARACTER_ in place of a variable list in the TABLES statement. However, when you use PROC TABULATE, you need to provide a list of variable names in the CLASS or VAR statement. Yes, it would be nice to use PROC TABULATE to provide a summary of missing values for all the variables in a SAS data set (and minimum and maximum values for the numeric variables) without having to "hard code" all of the variable names. A novel way to accomplish this task is to use PROC SQL to create two macro variables: one, a list of numeric variables in the data set; the other, a list of character variables. Program 3-8 uses this method, along with PROC TABULATE, to list the number of missing and nonmissing values for all the character and numeric variables in a data set.

Program 3-8   Writing a Macro to Count the Number of Missing and Nonmissing Observations for All Numeric and Character Variables in a Data Set

```
*-------------------------------------------------------------------*
| Program Name: AUTOMISS.SAS  in C:\CLEANING                        |
| Purpose: Macro to list the number of missing and nonmissing      |
|          variables in a SAS data set                             |
| Arguments: DSNAME =  SAS data set name (one- or two-level)       |
| Example: %AUTOMISS(CLEAN.PATIENTS)                               |
*-------------------------------------------------------------------*;

%MACRO AUTOMISS(DSNAME);

   %***One-level data set name;
   %IF %INDEX(&DSNAME,.) = 0 %THEN %DO;  ❶
      %LET LIB = WORK;
      %LET DSN = %UPCASE(&DSNAME);
   %END;

   %***Two-level data set name;
   %ELSE %DO;  ❷
      %LET LIB = %UPCASE(%SCAN(&DSNAME,1,"."));
      %LET DSN = %UPCASE(%SCAN(&DSNAME,2,"."));
   %END;

   %*Note: it is important for the libname and data set name to
     be in uppercase;

   %* Initialize macro variables to null;
   %LET NVARLIST=;
   %LET CVARLIST=;

   TITLE1 "Number of Missing and Nonmissing Values from &DSNAME";
   %* Get list of numeric variables;
   PROC SQL NOPRINT;
      SELECT NAME INTO :NVARLIST SEPARATED BY " "   ❸
      FROM DICTIONARY.COLUMNS   ❹
      WHERE LIBNAME = "&LIB" AND MEMNAME = "&DSN" AND TYPE = "num";  ❺

   %* Get list of character variables;
   SELECT NAME INTO :CVARLIST SEPARATED BY " "
      FROM DICTIONARY.COLUMNS
      WHERE LIBNAME = "&LIB" AND MEMNAME = "&DSN" AND TYPE = "char";  ❻
   QUIT;
```

```
   PROC FORMAT;  ❼
      VALUE $MISSCH " " = "Missing"
                   OTHER = "Nonmissing";
   RUN;


   PROC TABULATE DATA=&LIB..&DSN MISSING FORMAT=8.;   ❽

      %* If there are any numeric variables, do the following;
      %IF &NVARLIST NE %THEN %DO;
         VAR &NVARLIST;
         TITLE2 "for Numeric Variables";
         TABLE &NVARLIST,
            N NMISS MIN MAX / RTSPACE=26;
      %END;

      %* If there are any character variables, do the following;
      %IF &CVARLIST NE %THEN %DO;
         CLASS &CVARLIST;
         TITLE2 "for Character Variables";
         TABLE &CVARLIST,
            N / RTSPACE=26;
         FORMAT &CVARLIST $MISSCH.;
       %END;

      KEYLABEL N                      = "Number"
               NMISS                  = "Number Missing"
               MIN                    = "Lowest Value"
               MAX                    = "Highest Value";
   RUN;


%MEND AUTOMISS;
```

The macro starts with a test to see if the data set name (the calling argument) is a one- or two-level name. The macro function %INDEX ❶ returns a 0 if there is no period in the data set name. In this case, the macro variable LIB is set equal to WORK, and the macro variable DSN is set equal to the data set name. If the data set name contains a period, the %INDEX function returns a number greater than 0 and the two macro variables LIB and DSN are set equal to the libname and data set name, respectively ❷. It is important that the libname and data set names be in uppercase, so the %UPCASE function is used. The SELECT statement ❸ is used to place the variable name (NAME) in a macro variable called NVARLIST. This is accomplished by the terms INTO :NVARLIST. The list of names produced is separated by spaces, as indicated in the SELECT statement. The keyword DICTIONARY. COLUMNS ❹ returns the list of variables. Finally, to obtain the numeric and character lists separately, TYPE="num" or TYPE="char" is added to the WHERE statements ❺, ❻. When the two SELECT statements are processed, the macro variable NVARLIST will contain a list of all the numeric variables, separated by spaces and the macro variable CVARLIST will contain a list of all the character variables in the data set specified by the calling arguments.

The $MISSCH format ❼ is the same as we used in Program 3-7. Finally, the PROC TABULATE statements ❽ are identical to the statements in the previous programs where the variable lists were "hard coded," except the macro variables are substituted for the explicit list of variables.

Calling this macro with the data set name CLEAN.PATIENTS as the argument, %AUTOMISS(CLEAN.PATIENTS), produced the following output:

Number of Missing and Nonmissing Values from CLEAN.PATIENTS for Numeric Variables

	Number	Number Missing	Lowest Value	Highest Value
Visit Date	24	7	13966	14560
Heart Rate	28	3	10	900
Systolic Blood Pressure	27	4	20	400
Diastolic Blood Pressure	28	3	8	200

Number of Missing and Nonmissing Values from CLEAN.PATIENTS for Character Variables

	Number
Patient Number	
Missing	1
Nonmissing	30
Gender	
Missing	1
Nonmissing	30
Diagnosis Code	
Missing	8
Nonmissing	23
Adverse Event?	
Missing	1
Nonmissing	30

The PROC MEANS and PROC FREQ methods for listing missing values are certainly easier than the complicated program you were shown before. You may prefer the appearance of the PROC TABULATE output better, or you may just be interested in the programming techniques that were used.

Searching for a Specific Numeric Value

Specific values such as 999 or 9999 are sometimes used to denote missing values. For example, numeric values that are left blank in Dbase files are stored as zeros. If 0 is a valid value for some of your variables, this can lead to problems. So, values such as 999 or 9999 are sometimes used instead of zeros or blanks. Program 3-9 searches a SAS data set for all numeric variables set to a specific value and produces a report, which shows the variable name and the observation where the specific value was found.

The "trick" in this program is the relatively unknown routine, VNAME (see online help for SAS Release 6.12 or later for more details). A call to this routine returns the variable name of an array element. The first program (Program 3-9) searches a SAS data set for a specific value. The program is then generalized by making the data set name and the specific value, calling arguments in a macro. Here is the first program.

Program 3-9   Identifying All Numeric Variables Equal to a Fixed Value (Such as 999)

```
*-----------------------------------------------------------------*
| Program Name: FIND_X.SAS   in C:\CLEANING                       |
| Purpose: Identifies any specified value for all numeric vars    |
*-----------------------------------------------------------------*;
***Create test data set;
DATA TEST;
   INPUT X Y A $ X1-X3 Z $;
DATALINES;
1 2 X 3 4 5 Y
2 999 Y 999 1 999 J
999 999 R 999 999 999 X
1 2 3 4 5 6 7
;
```

```
***Program to detect the specified values;
DATA _NULL_;
   SET TEST;
   FILE PRINT;
   ARRAY NUMS[*] _NUMERIC_;    ❶
   LENGTH VARNAME $ 8;
   DO __I = 1 TO DIM(NUMS);    ❷
      IF NUMS[__I] = 999 THEN DO;
         CALL VNAME(NUMS[__I],VARNAME);    ❸
         PUT "Value of 999 found for variable " VARNAME
            "in observation " _N_;
      END;
   END;
   DROP __I;
RUN;
```

Key to this program is the use of _NUMERIC_ in the ARRAY statement ❶. Because this ARRAY statement follows the SET statement, the array NUMS will contain all the numeric variables in the data set TEST. The next step is to examine each of the elements in the NUMS array, determine if a value of 999 is found, and then determine the variable name associated with that array element. The DO loop ❷ uses the index variable __I in the hopes that there will not be any variables in the data set to be tested with that name.

Now for the "trick." As you search for values of 999 for each of the numeric variables, you can use the CALL VNAME routine ❸ (see the online help for Release 6.12 or later) to return the variable name that corresponds to the array element. In this program, the variable name is stored in the variable VARNAME (the first argument) by the VNAME routine. All that is left to do is write out the variable names and observation numbers. Next is the macro version of the same program, followed by the output.

Program 3-10 Creating a Macro Version of Program 3-9

```
*------------------------------------------------------------------*
| Macro Name: FIND_X.SAS  in C:\CLEANING                           |
| Purpose: Identifies any specified value for all numeric vars     |
| Calling Arguments: DSN    SAS Data Set Name                      |
|                    NUM    Numeric value to search for            |
| Example:  To find variable values of 999 in data set TEST, use   |
|           %FIND_X(TEST,999)                                       |
*------------------------------------------------------------------*;
```

```
%MACRO FIND_X(DSN,NUM);
   TITLE "Variables with 999 as Missing Values";
   DATA _NULL_;
      SET &DSN;
      FILE PRINT;
      LENGTH VARNAME $ 8; ***Or LENGTH 32 for V7 and Later;
      ARRAY NUMS[*] _NUMERIC_;
      DO __I = 1 TO DIM(NUMS);
         IF NUMS[__I] = &NUM THEN DO;
            CALL VNAME(NUMS[__I],VARNAME);
            PUT "Value of &NUM found for variable " VARNAME
               "in observation " _N_;
         END;
      END;
      DROP __I;
   RUN;
%MEND FIND_X;
```

You call this macro by using the following statement:

```
%FIND_X(TEST,999)
```

The resulting output is shown next.

```
Variables with 999 as Missing Values

Value of 999 found for variable Y in observation 2
Value of 999 found for variable X1 in observation 2
Value of 999 found for variable X3 in observation 2
Value of 999 found for variable X in observation 3
Value of 999 found for variable Y in observation 3
Value of 999 found for variable X1 in observation 3
Value of 999 found for variable X2 in observation 3
Value of 999 found for variable X3 in observation 3
```

If you would prefer just to see a summary of variables that have the value of a number, such as 999, for one or more observations, you can modify Program 3-9 to create a data set, and use PROC FREQ to count the number of times a specified value is detected as shown in Program 3-11.

Program 3-11 Identifying Variables with Specified Numeric Values and Counting the Number of Times the Value Appears

```
DATA NUM_999;
   SET TEST;
   FILE PRINT;
   ARRAY NUMS[*] _NUMERIC_;
   LENGTH VARNAME $ 8;
   DO __I = 1 TO DIM(NUMS);
      IF NUMS[__I] = 999 THEN DO;
         CALL VNAME(NUMS[__I],VARNAME);
         OUTPUT;
      END;

   END;
   KEEP VARNAME;
RUN;

PROC FREQ DATA=NUM_999;
   TABLES VARNAME / NOCUM NOPERCENT;
RUN;
```

Each time a numeric variable (in the array NUMS) is equal to a value of 999, a call to VNAME places the variable name into the variable VARNAME. An observation is then written to the data set NUM_999. Because you want to count the number of times the specific numeric value (such as 999) occurred, use PROC FREQ to print out the frequencies.

Running Program 3-11 with the value of 999 on the data set TEST generated the following output.

```
Variables with 999 as Missing Values

The FREQ Procedure

VARNAME      Frequency
---------------------
X                   1
X1                  2
X2                  1
X3                  2
Y                   2
```

You can convert this last program into a SAS macro as an exercise.

4 Working with Dates

Introduction

SAS dates seem mysterious to many people, but by understanding how SAS dates are stored, you will see that they are really quite simple. SAS dates are stored in numeric variables and represent the number of days from a fixed point in time, January 1, 1960. The confusion develops in the many ways that SAS software can read and write dates. Typically, dates are read as MM/DD/YYYY or some similar form. There are informats to read almost any conceivable date notation. Regardless of how a date is read, the informat performs the conversion to a SAS date, and it is stored just like any other numeric value. If you print out a date value without a SAS date format, it will appear as a number (the number of days from January 1, 1960) rather than a date in one of the standard forms. When date information is not in a standard form, you can read the month, day, and year information as separate variables and use the MDY (month-day-year) function to create a SAS date. Let's look at some ways to perform data cleaning and validation with dates.

Checking Ranges for Dates (Using a DATA Step)

Suppose you want to determine if the visit dates in the PATIENTS data set are between June 1, 1998 and October 15, 1999. You can use a DATA step approach, much the same way you did when checking that numeric variables were within a specified range. The only difference is that the range boundaries have to be SAS dates. Let's see how this works. Program 4-1 checks for dates in the specified range and ignores missing values.

Program 4-1   Checking That a Date Is within a Specified Interval (DATA Step Approach)

```
LIBNAME CLEAN "C:\CLEANING";

DATA _NULL_;
   TITLE "Dates before June 1, 1998 or after October 15, 1999";
   FILE PRINT;
   SET CLEAN.PATIENTS(KEEP=VISIT PATNO);
   IF VISIT LT '01JUN1998'D AND VISIT NE . OR
      VISIT GT '15OCT1999'D THEN PUT PATNO= VISIT= MMDDYY10.;
RUN;
```

The key to this program is the use of the date constants (also called date literals) in the IF statement. If you want SAS to turn a date into a SAS date (the number of days from 1/1/1960), the dates must be written in this fashion. Date constants are written as a two-digit day, a three-character month name, and a two- or four-digit year, placed in single or double quotes, and followed by a lowercase or uppercase 'D'. You also need to add a date format in the PUT statement so that the date will be printed in a standard date format. Output from Program 4-1 is shown next.

```
Dates before June 1, 1998 or after October 15, 1999

PATNO=XX5 VISIT=05/07/1998
PATNO=010 VISIT=10/19/1999
PATNO=003 VISIT=11/12/1999
PATNO=028 VISIT=03/28/1998
PATNO=029 VISIT=05/15/1998
```

Checking Ranges for Dates (Using PROC PRINT)

You can accomplish the same objective by using a PROC PRINT statement in combination with a WHERE statement. Besides being easier to code, this approach allows you to use the keyword BETWEEN with the WHERE statement, making the logic somewhat simpler to follow. The SAS code is shown in Program 4-2.

Program 4-2   Checking That a Date Is within a Specified Interval (Using PROC PRINT and a WHERE Statement)

```
PROC PRINT DATA=CLEAN.PATIENTS;
   TITLE "Dates before June 1, 1998 or after October 15, 1999";
   WHERE VISIT NOT BETWEEN '01JUN1998'D AND '15OCT1999'D AND VISIT NE .;
   ID PATNO;
   VAR VISIT;
   FORMAT VISIT DATE9.;
RUN;
```

Output from this procedure contains the identical information as the previous DATA step approach. For variety, let's choose the DATE9. date format.

```
Dates before June 1, 1998 or after October 15, 1999

PATNO        VISIT

  XX5      07MAY1998
  010      19OCT1999
  003      12NOV1999
  028      28MAR1998
  029      15MAY1998
```

Checking for Invalid Dates

Some of the dates in the PATIENTS data set are missing and some are invalid dates, which were converted to missing values during the input process. If you want to distinguish between the two, you must work from the raw data, not the SAS data set. If you attempt to read an invalid date with a SAS date informat, an error message will appear in the SAS Log. This is one clue that you have errors in your date values. Program 4-3 reads the raw data file PATIENTS.TXT. The resulting SAS Log follows.

Program 4-3   Reading Dates with the MMDDYY10. Informat

```
DATA DATES;
   INFILE "C:\CLEANING\PATIENTS.TXT" PAD;
   INPUT @5 VISIT MMDDYY10.;
   FORMAT VISIT MMDDYY10.;
RUN;
```

The SAS Log that results from running this program is shown next.

```
1     LIBNAME CLEAN "C:\CLEANING";
NOTE: Libref CLEAN was successfully assigned as follows:
      Engine:         V6
      Physical Name: C:\CLEANING
2     DATA DATES;
3        INFILE "C:\CLEANING\PATIENTS.TXT" PAD;
4        INPUT @5 VISIT MMDDYY10.;
5        FORMAT VISIT MMDDYY10.;
6     RUN;

NOTE: The infile "C:\CLEANING\PATIENTS.TXT" is:
      File Name=C:\CLEANING\PATIENTS.TXT,
      RECFM=V,LRECL=256

NOTE: Invalid data for VISIT in line 7 5-14.
RULE:    ---+---1---+---2---+---3---+---4---+---5---+---6---+---7---+---8
7        007M08/32/1998 88148102    0
      87
     173
VISIT=. _ERROR_=1 _N_=7
NOTE: Invalid data for VISIT in line 12 5-14.
12        011M13/13/1998 68300 20   41
      87
     173
VISIT=. _ERROR_=1 _N_=12
NOTE: Invalid data for VISIT in line 21 5-14.
21        123M15/12/1999 60        10
      87
     173
VISIT=. _ERROR_=1 _N_=21
NOTE: Invalid data for VISIT in line 23 5-14.
23        020F99/99/9999 10 20  8   0
      87
     173

VISIT=. ERROR_=1 _N_=23
```

continued

```
NOTE: Invalid data for VISIT in line 28 5-14.
28         027FNOTAVAIL   NA 166106   70
      87
     173
VISIT=. _ERROR_=1 _N_=28
NOTE: 31 records were read from the infile "C:\CLEANING\PATIENTS.TXT".
      The minimum record length was 26.
      The maximum record length was 27.
NOTE: The data set WORK.DATES has 31 observations and 1 variables.
NOTE: DATA statement used:
      real time            2.14 seconds
```

There are several reasons why these dates caused error reports in the Log. In some cases, the month was greater than 12, possibly caused by reading a date that was actually in day-month-year form rather than month-day-year form. Other dates such as 99/99/9999 were an attempt to indicate that no date information was available. Obviously, the data value of NOTAVAIL (not available) caused an error. Remember that once the errors exceed a default number, they will no longer be reported in the SAS Log. This number can be adjusted by setting the system option ERRORS=. If you have no missing date values in your data, any missing date value must have been generated by an invalid date. You can use this idea to list missing and invalid dates. The plan is to read the date value twice; once with a SAS date informat, the other as a character string. This way, you can see the original date value that caused the error. Program 4-4 demonstrates this.

Program 4-4   Listing Missing and Invalid Dates by Reading the Date Twice, Once with a Date Informat and the Second as Character Data

```
DATA _NULL_;
   FILE PRINT;
   TITLE "Listing of Missing and Invalid Dates";
   INFILE "C:\CLEANING\PATIENTS.TXT" PAD;
   INPUT @1 PATNO $3.
         @5 VISIT MMDDYY10.
         @5 V_DATE $CHAR10.;
   FORMAT VISIT MMDDYY10.;
   IF VISIT = . THEN PUT PATNO= V_DATE=;
RUN;
```

Here you read the date twice, first with the SAS date informat MMDDYY10. and then with the character informat $CHAR10. This way, even though the SAS System substitutes a missing value for VISIT, the variable V_DATE will contain the actual characters that were entered in the date field.

An alternative is to read the original date only once as a character string, and use the INPUT function to create the SAS date. Remember that the INPUT function "reads" the value of the first argument, using the informat listed as the second argument. The alternative program is shown in Program 4-5.

Program 4-5   Listing Missing and Invalid Dates by Reading the Date as a Character Variable and Converting to a SAS Date with the INPUT Function

```
DATA _NULL_;
   FILE PRINT;
   TITLE "Listing of Missing and Invalid Dates";
   INFILE "C:\CLEANING\PATIENTS.TXT" PAD;
   INPUT @1 PATNO $3.
         @5 V_DATE $CHAR10.;
   VISIT = INPUT(V_DATE,MMDDYY10.);
   FORMAT VISIT MMDDYY10.;
   IF VISIT = . THEN PUT PATNO= V_DATE=;
RUN;
```

Running either Program 4-4 or Program 4-5 results in the following output.

```
Listing of Missing and Invalid Dates

PATNO=007 V_DATE=08/32/1998
PATNO=011 V_DATE=13/13/1998
PATNO=015 V_DATE=
PATNO=123 V_DATE=15/12/1999
PATNO=321 V_DATE=
PATNO=020 V_DATE=99/99/9999
PATNO=027 V_DATE=NOTAVAIL
```

If you want to ignore real missing values, you only need to make a slight change as shown in Program 4-6.

Program 4-6   Removing the Missing Values from the Invalid Date Listing

```
DATA _NULL_;
   FILE PRINT;
   INFILE "C:\CLEANING\PATIENTS.TXT" PAD;
   INPUT @1 PATNO $3.
         @5 V_DATE $CHAR10.;
```

```
   VISIT = INPUT(V_DATE,MMDDYY10.);
   FORMAT VISIT MMDDYY10.;
   IF VISIT = . AND V_DATE NE ' ' THEN PUT PATNO= V_DATE=;    ❶
RUN;
```

Because of line ❶, only nonmissing, invalid dates will be printed.

Working with Dates in Nonstandard Form

Although SAS software can read dates in almost every conceivable form, there may be times when you have date information for which there is no SAS informat. Suppose you have a month value (a number from 1 to 12) in columns 6-7, a day of the month value (a number from 1 to 31) in columns 13-14, and a four-digit year value in columns 20-23. How can you create a SAS date from these three variables? The MDY (month-day-year) function comes to the rescue. Just enter the three variable names for the month, day, and year as arguments to this function, and it will return a SAS date. Program 4-7 demonstrates how this works.

Program 4-7 Demonstrating the MDY Function to Read Dates in Nonstandard Form

```
***Sample program to read nonstandard dates;
DATA NONSTAND;
   INPUT PATNO $ 1-3 MONTH 6-7 DAY 13-14 YEAR 20-23;
   DATE = MDY(MONTH,DAY,YEAR);
   FORMAT DATE MMDDYY10.;
DATALINES;
001  05     23      1998
006  11     01      1998
123  14     03      1998
137  10             1946
;
PROC PRINT DATA=NONSTAND;
   TITLE "Listing of Data Set NONSTAND";
   ID PATNO;
RUN;
```

Notice that an invalid MONTH value (observation three) and a missing DAY value (observation four) were included intentionally. The listing of the data set NONSTAND follows.

```
Listing of Data Set NONSTAND

PATNO    MONTH    DAY    YEAR        DATE

 001       5       23    1998    05/23/1998
 006      11        1    1998    11/01/1998
 123      14        3    1998        .
 137      10        .    1946        .
```

In the two cases where a date could not be computed, a missing value was generated. Inspection of the SAS Log also shows that the MDY function had an invalid value and a missing value. You can check for invalid dates in the same way you did earlier. You can create a character string representing the date in a standard date format such as MMDDYY10. and test for missing values. Program 4-8 flags invalid nonmissing dates.

Program 4-8   Removing Missing Values from the Error Listing

```
DATA _NULL_;
   FILE PRINT;
   TITLE "Invalid Date Values";
   INPUT PATNO $ 1-3 MONTH 6-7 DAY 13-14 YEAR 20-23;
   DATE = MDY(MONTH,DAY,YEAR);
   C_DATE = PUT(MONTH,Z2.) || '/' ||
            PUT(DAY,Z2.) || '/' ||
            PUT(YEAR,4.);
   ***Note: the Z2. Format includes leading zeros;
   FORMAT DATE MMDDYY10.;
   IF C_DATE NE ' ' AND DATE = . THEN PUT PATNO= C_DATE=;
DATALINES;
001  05     23     1998
006  11     01     1998
123  14     03     1998
137  10            1946
;
```

In Program 4-8, C_DATE is a character representation of the date. The concatenation operator, ||, is used to piece together the month, day, and year values and the two slashes. If the MDY function produces a missing value and the value of C_DATE is nonmissing, there must have been an invalid date. The output that follows demonstrates that this program works as advertised.

```
Invalid Date Values

PATNO=123 C_DATE=14/ 3/1998
PATNO=137 C_DATE=10/ ./1946
```

Creating a SAS Date When the Day of the Month Is Missing

Some of your date values may be missing the day of the month, but you would still like to create a SAS date by using either the 1st or the 15th of the month as the day. There are two possibilities here. One method is to use the MONYY informat that reads dates in the form of a three-character month name and a two- or four-digit year. If your dates are in this form, SAS will create a SAS date using the first of the month as the day value. The other method of creating a SAS date from only month and year values is to use the MDY function, substituting a value such as 15, for the day argument. An example is shown in Program 4-9.

Program 4-9   Creating a SAS Date When the Day of the Month Is Missing

```
DATA NO_DAY;
   INPUT @1 DATE1 MONYY7. @8 MONTH 2. @10 YEAR 4.;
   DATE2 = MDY(MONTH,15,YEAR);
   FORMAT DATE1 DATE2 MMDDYY10.;
DATALINES;
JAN98  011998
OCT1998101998
;
PROC PRINT DATA=NO_DAY;
   TITLE "Listing of Data Set NO_DAY";
RUN;
```

DATE1 is a SAS date created by the MONYY SAS informat; DATE2 is created by the MDY function, using the 15th of the month as the missing day value. Output from PROC PRINT is shown next.

```
Listing of Data Set NO_DAY

Obs        DATE1      MONTH     YEAR        DATE2

 1      01/01/1998       1      1998     01/15/1998
 2      10/01/1998      10      1998     10/15/1998
```

Let's extend this idea a bit further. Suppose most of your dates have month, day, and year values but, for any date where the only piece missing is the day of the month, you want to substitute the 15th of the month. Program 4-10 will accomplish this goal.

Program 4-10 Substituting the 15th of the Month When the Date of the Month Is Missing

```
DATA MISS_DAY;
   INPUT @1  PATNO   $3.
         @4  MONTH   2.
         @6  DAY     2.
         @8  YEAR    4.;
   IF DAY NE . THEN DATE = MDY(MONTH,DAY,YEAR);
   ELSE DATE = MDY(MONTH,15,YEAR);
   FORMAT DATE MMDDYY10.;
DATALINES;
00110211998
00205  1998
00344  1998
;
PROC PRINT DATA=MISS_DAY;
   TITLE "Listing of Data Set MISS_DAY";
RUN;
```

If the day value is not missing, the MDY function uses all three values of month, day, and year to compute a SAS date. If the day value is missing, the 15th of the month is used. As before, if there is an invalid date (such as for patient 003), a missing date value is generated. Here are the three observations created by this program.

```
Listing of Data Set MISS_DAY

OBS    PATNO    MONTH    DAY    YEAR        DATE

 1      001      10      21     1998    10/21/1998
 2      002       5       .     1998    05/15/1998
 3      003      44       .     1998       .
```

Suspending Error Checking for Known Invalid Dates

As you saw earlier, invalid date values can fill your SAS Log with lots of errors. There are times when you know that invalid date values were used to represent missing dates or other specific values. If you would like to prevent the automatic listing of date errors in the SAS Log, you can use the double question mark (??) modifier in your INPUT statement or with the INPUT function. This modifier prevents the NOTES and data listings to be printed in the SAS Log and also keeps the SAS internal variable _ERROR_ at 0.

Program 4-11 uses the ?? modifier in the INPUT statement to prevent error messages from printing in the SAS Log.

Program 4-11 Suspending Error Checking for Known Invalid Dates by Using the ?? Informat Modifier

```
DATA DATES;
   INFILE "C:\CLEANING\PATIENTS.TXT" PAD;
   INPUT @5 VISIT ?? MMDDYY10.;
   FORMAT VISIT MMDDYY10.;
RUN;
```

When this program is run, there will be no error messages in the SAS Log caused by invalid dates. Only turn off SAS error checking when you plan to detect errors in other ways, or you already know all about your invalid dates.

Program 4-12 shows an example of using the ?? informat modifier with the INPUT function. The following program is identical to Program 4-5, with the addition of the ?? modifier to keep the SAS Log free of error messages.

Program 4-12 Demonstrating the ?? Informat Modifier with the INPUT Function

```
DATA _NULL_;
   FILE PRINT;
   INFILE "C:\CLEANING\PATIENTS.TXT" PAD;
   INPUT @1 PATNO $3.
         @5 V_DATE $CHAR10.;
   VISIT = INPUT(V_DATE,?? MMDDYY10.);
   FORMAT VISIT MMDDYY10.;
   IF VISIT = . THEN PUT PATNO= V_DATE=;
RUN;
```

Remember that you can use the ?? modifier before the informat argument of the INPUT function as well as the more traditional use with the INPUT statement. If you have a lot of known date errors, the overriding of the error messages will also improve program efficiency. (Remember that this also sets the value of _ERROR_ to 0.)

5

Looking for Duplicates and "n" Observations per Subject

Introduction

Besides checking for invalid data values in a data set, it may be necessary to check for either duplicate ID's or duplicate observations. Duplicate observations are easy to fix; just eliminate the duplicates (although you may want to find out how the duplicates got there). Duplicate ID's with different data values presents another problem. One possible cause of this is that the same ID was used for more than one person. Another possibility is that different data values were entered more than once for the same person. There are several ways to detect and eliminate duplicates in a SAS data set. This chapter explores some of them.

Eliminating Duplicates by Using PROC SORT

Suppose you have a data set where each patient is supposed to be represented by a single observation. To demonstrate what happens when you have multiple observations with the same ID, some duplicates in the PATIENTS data set were included on purpose. Observations with duplicate ID numbers are shown next.

OBS	PATNO	GENDER	VISIT	HR	SBP	DBP	DX	AE
2	002	F	11/13/1998	84	120	78	X	0
3	002	F	11/13/1998	84	120	78	X	0
4	003	X	10/21/1998	68	190	100	3	1
5	003	M	11/12/1999	58	112	74		0
7	006		06/15/1999	72	102	68	6	1
8	006	F	07/07/1999	82	148	84	1	0

Notice that patient number 002 is a true duplicate observation. For patient numbers 003 and 006, the duplicate ID's contain different values.

Two very useful options of PROC SORT are NODUPKEY and NODUP. The NODUPKEY option automatically eliminates multiple observations where the BY variables have the same value. For example, to automatically eliminate multiple patient ID's (PATNO) in the PATIENTS data set (which you probably would not want to do, this is for illustration only), you could use PROC SORT with the NODUPKEY option as shown in Program 5-1.

Program 5-1   Demonstrating the NODUPKEY Option of PROC SORT

```
PROC SORT DATA=CLEAN.PATIENTS OUT=SINGLE NODUPKEY;
   BY PATNO;
RUN;

PROC PRINT DATA=SINGLE;
   TITLE "Data Set SINGLE - Duplicated ID's Removed from PATIENTS";
   ID PATNO;
RUN;
```

Notice that two options, OUT= and NODUPKEY, are used here. The OUT= option is used to create the new data set SINGLE, leaving the original data set PATIENTS unchanged. Shown next is a listing of the SINGLE data set.

```
Data Set SINGLE - Duplicated ID's Removed from PATIENTS

PATNO     GENDER         VISIT      HR     SBP    DBP     DX    AE

            M         11/11/1998    90     190    100           0
 001        M         11/11/1998    88     140     80     1     0
 002        F         11/13/1998    84     120     78     X     0
 003        X         10/21/1998    68     190    100     3     1
 004        F         01/01/1999   101     200    120     5     A
 006                  06/15/1999    72     102     68     6     1
 007        M             .         88     148    102           0
 008        F         08/08/1998   210      .      .      7     0
 009        M         09/25/1999    86     240    180     4     1
 010        f         10/19/1999     .      40    120     1     0
 011        M             .         68     300     20     4     1
 012        M         10/12/1998    60     122     74           0
 013        2         08/23/1999    74     108     64     1
 014        M         02/02/1999    22     130     90           1
 015        F             .         82     148     88     3     1
 017        F         04/05/1999   208      .      84     2     0
 019        M         06/07/1999    58     118     70           0
 020        F             .         10      20      8           0
 022        M         10/10/1999    48     114     82     2     1
 023        f         12/31/1998    22      34     78           0
 024        F         11/09/1998    76     120     80     1     0
 025        M         01/01/1999    74     102     68     5     1
 027        F             .          .     166    106     7     0
 028        F         03/28/1998    66     150     90     3     0
 029        M         05/15/1998     .      .      .      4     1
 123        M             .         60      .      .      1     0
 321        F             .        900     400    200     5     1
 XX5        M         05/07/1998    68     120     80     1     0
```

The NODUPKEY option eliminated the second observation for each of the three duplicate ID's. The only indication that duplicates were removed is in the NOTE in the SAS Log, which is shown next.

```
507   PROC SORT DATA=CLEAN.PATIENTS OUT=SINGLE NODUPKEY;
508      BY PATNO;
509   RUN;

NOTE: 3 observations with duplicate key values were deleted.
NOTE: The data set WORK.SINGLE has 28 observations and 8 variables.
```

This method of looking for duplicate ID's is really only useful if the SAS Log shows that no duplicates were removed. If the SAS Log shows duplicate key values were deleted, you need to see which ID's had duplicate data and the nature of the data.

If you use the NODUPKEY option with more than one BY variable, only those observations with identical values on each of the BY variables will be deleted. For example, if you sort by patient number (PATNO) and visit date (VISIT), only the duplicate for patient number 002 will be deleted when you use the NODUPKEY option because the two observations for patient number 002 are the only ones with the same patient number and visit date.

The option NODUP also deletes duplicates, but only for two observations where all the variables have identical values. Program 5-2 demonstrates this option.

Program 5-2   Demonstrating the NODUP Option of PROC SORT

```
PROC SORT DATA=CLEAN.PATIENTS OUT=SINGLE NODUP;
   BY _ALL_;
RUN;
```

Listing the data set SINGLE, which is created by this procedure, shows that only the second observation for patient number 002 was deleted. The use of _ALL_ as the BY variable in Program 5-2 is necessary because of the strange way that the NODUP option looks for duplicates. PROC SORT compares an observation to the most recently written observation when deciding if an observation is a duplicate, so you have to sort by all the variables to make it work. (This piece of information is due to one of my very astute reviewers, Mike Zdeb, who demonstrated that simply sorting by a single variable and using the NODUP option could result in a data set that still had duplicate observations!)

Because this is such an important point, Program 5-3 illustrates how the NODUP option can leave duplicate observations in a data set when you do not use the keyword _ALL_ in the BY statement.

Program 5-3   Demonstrating a "Feature" of the NODUP Option

```
DATA MULTIPLE;
   INPUT PATNO $ X Y;
DATALINES;
001 1 2
006 1 2
009 1 2
001 3 4
001 1 2
009 1 2
001 1 2
;
PROC SORT DATA=MULTIPLE OUT=SINGLE NODUP;
   BY PATNO;
RUN;
PROC PRINT DATA=SINGLE;
   TITLE "Listing of Data Set SINGLE";
RUN;
```

When PROC SORT sorts the observations by the variable PATNO, the four observations with PATNO equal to '001' wind up as the first four observations in the sorted data set. The first four observations in the sorted data set are

```
001 1 2
001 3 4
001 1 2
001 1 2
```

Now, when the NODUP option "does its thing," successive duplicate observations are deleted, leaving the first three observations for PATNO '001' in the data set SINGLE. You can see this in the listing of the data set SINGLE, which is shown next.

```
Listing of Data Set SINGLE

Obs      PATNO      X      Y
 1        001       1      2
 2        001       3      4
 3        001       1      2
 4        006       1      2
 5        009       1      2
```

This is an important point that you must keep in mind whenever you want to eliminate duplicate observations by using the NODUP option of PROC SORT.

Detecting Duplicates by Using DATA Step Approaches

Let's explore the ways that will allow you to detect duplicate ID's and duplicate observations in a data set. One very good way to approach this problem is to use the temporary SAS variables FIRST. and LAST. To see how this works, look at Program 5-4, which prints out all observations that have duplicate patient numbers.

Program 5-4   Identifying Duplicate ID's

```
PROC SORT DATA=CLEAN.PATIENTS OUT=TMP;    ❶
   BY PATNO;
RUN;

DATA DUP;
   SET TMP;
   BY PATNO;   ❷
   IF FIRST.PATNO AND LAST.PATNO THEN DELETE;   ❸
RUN;

PROC PRINT DATA=DUP;
   TITLE "Listing of Duplicates from Data Set CLEAN.PATIENTS";
   ID PATNO;
RUN;
```

It's first necessary to sort the data set by the ID variable ❶. In the above program, the original data set was left intact and a new data set (TMP) was created for the sorted observations. After you have a sorted data set, a short DATA step will remove patients that have a single observation, leaving a data set of duplicates. The key here is the BY statement ❷ following the SET statement. When a SET statement is followed by a BY statement, the temporary SAS variables FIRST.*by_variable_name* and LAST.*by _variable _name* are created. In this example, there is only one BY variable (PATNO) so the two temporary SAS variables FIRST.PATNO and LAST.PATNO are created. If an observation is the first one in a BY group (in this case, the first occurrence of a patient number), FIRST.PATNO will be true (equal to one). If an observation is the last one in a BY group, LAST.PATNO will be true. Obviously, if FIRST.PATNO and LAST.PATNO are both true, there is only one observation for that patient number ❸.

Therefore, the data set DUP contains only observations where there is more than one observation for each patient number. Output from Program 5-4 is shown next.

```
Listing of Duplicates from Data Set CLEAN.PATIENTS

PATNO     GENDER         VISIT      HR    SBP    DBP    DX    AE

 002        F        11/13/1998     84    120    78     X     0
 002        F        11/13/1998     84    120    78     X     0
 003        X        10/21/1998     68    190    100    3     1
 003        M        11/12/1999     58    112    74           0
 006                 06/15/1999     72    102    68     6     1
 006        F        07/07/1999     82    148    84     1     0
```

Next, you want to consider the case where there are two or more observations for each patient, and each observation is supposed to have a different visit date (VISIT). The data set PATIENTS2 was created to demonstrate this situation. For this data set, two observations with the same patient ID and visit date would constitute an error. Refer to the Appendix for a listing of the raw data file PATIENTS2.TXT from which the PATIENTS2 data set was created.

If you would like to create the PATIENTS2 data set for test purposes, run the short DATA step shown in Program 5-5.

Program 5-5   Creating the SAS Data Set PATIENTS2 (a Data Set Containing Multiple Visits for Each Patient)

```
LIBNAME CLEAN "C:\CLEANING";

DATA CLEAN.PATIENTS2;
   INFILE "C:\CLEANING\PATIENTS2.TXT" PAD;
   INPUT @1  PATNO  $3.
         @4  VISIT  MMDDYY10.
         @14 HR      3.
         @17 SBP     3.
         @20 DBP     3.;
   FORMAT VISIT MMDDYY10.;
RUN;
```

A listing of the resulting data set from Program 5-5 follows.

```
Listing of Data set PATIENTS2

OBS    PATNO       VISIT      HR    SBP    DBP

 1      001     06/12/1998    80    130    80
 2      001     06/15/1998    78    128    78
 3      002     01/01/1999    48    102    66
 4      002     01/10/1999    70    112    82
 5      002     02/09/1999    74    118    78
 6      003     10/21/1998    68    120    70
 7      004     03/12/1998    70    102    66
 8      004     03/13/1998    70    106    68
 9      005     04/14/1998    72    118    74
10      005     04/14/1998    74    120    80
11      006     11/11/1998   100    180   110
12      007     09/01/1998    68    138   100
13      007     10/01/1998    68    140    98
```

Notice that there are from one to three observations for each patient. Also, notice that patient 005 has two observations with the same VISIT date and different data values. To detect this situation, use the variables FIRST. and LAST., except with two BY variables instead of one, as shown in Program 5-6.

Program 5-6   Identifying Patient ID's with Duplicate Visit Dates

```
PROC SORT DATA=CLEAN.PATIENTS2 OUT=TMP;   ❶
   BY PATNO VISIT;
RUN;

DATA DUP;
   SET TMP;
   BY PATNO VISIT;   ❷
   IF FIRST.VISIT AND LAST.VISIT THEN DELETE;   ❸
RUN;

PROC PRINT DATA=DUP;
   TITLE "Listing of Duplicates from Data Set CLEAN.PATIENTS2";
   ID PATNO;
RUN;
```

You sort as before ❶, only this time, execute a two-level sort, by PATNO and VISIT ❷. This places the observations in the data set TMP in patient number and visit date order. The SAS temporary variable FIRST.VISIT will be true for the first unique patient number and visit date combination ❸. Only the two identical visit dates for patient 005 will be selected for the data set DUP as demonstrated in the listing shown next.

```
Listing of Duplicates from Data Set CLEAN.PATIENTS2

PATNO         VISIT      HR     SBP     DBP

  005      04/14/1998    72     118      74
  005      04/14/1998    74     120      80
```

Using PROC FREQ to Detect Duplicate ID's

Another way to find duplicates uses PROC FREQ to count the number of observations for each value of the patient ID variable (PATNO). Use the patient ID variable and the OUT= option in the TABLES statement to create a SAS data set that contains the value of PATNO and the frequency count (PROC FREQ uses the variable name COUNT to hold the frequency information). After you have this information, you can use it to select the original duplicate observations from your data set. To demonstrate how this works, Program 5-7 identifies duplicate patient numbers from the PATIENTS data set.

Program 5-7   Using PROC FREQ and an Output Data Set to Identify Duplicate ID's

```
PROC FREQ DATA=CLEAN.PATIENTS NOPRINT;    ❶
   TABLES PATNO / OUT=DUP_NO(KEEP=PATNO COUNT
                        WHERE=(COUNT GT 1));    ❷
RUN;

PROC SORT DATA=CLEAN.PATIENTS OUT=TMP;
   BY PATNO;
RUN;

PROC SORT DATA=DUP_NO;
   BY PATNO;
RUN;
```

```
DATA DUP;
   MERGE TMP DUP_NO(IN=YES_DUP DROP=COUNT);   ❸
   BY PATNO;
   IF YES_DUP;   ❹
RUN;

PROC PRINT DATA=DUP;
   TITLE "Listing of Data Set DUP";
RUN;
```

PROC FREQ ❶ uses the NOPRINT option because you only want the output data set, not the actual PROC FREQ listing. The OUT= option in the TABLES statement ❷ creates a SAS data set called DUP_NO, which contains the variables PATNO and COUNT. The WHERE data set option restricts this data set to those observations where COUNT is greater than one (the duplicates).

Next, sort both the original data set PATIENTS (to the temporary data set TMP) and the DUP_NO data set by PATNO. The final DATA step merges the two data sets. The key to the entire program is the IN= option in this MERGE statement ❸. The DUP_NO data set only contains patient numbers where the value of COUNT is greater than one. The logical variable YES_DUP, created by this IN= data set option, is true whenever the DUP_NO data set is making a contribution to the observation being formed. Thus, because of line ❹, only the duplicates will be placed in the DUP data set, as shown in the next listing.

```
Listing of Data Set DUP

OBS    PATNO    GENDER      VISIT       HR    SBP    DBP    DX    AE

  1     002      F        11/13/1998    84    120    78     X     0
  2     002      F        11/13/1998    84    120    78     X     0
  3     003      X        10/21/1998    68    190    100    3     1
  4     003      M        11/12/1999    58    112    74           0
  5     006               06/15/1999    72    102    68     6     1
  6     006      F        07/07/1999    82    148    84     1     0
```

If all you need to do is identify the patient ID's with more than one observation, you can avoid the MERGE step because the output data set from PROC FREQ contains the variable PATNO as well as the frequency (COUNT). So, the much simpler program is shown in Program 5-8.

Program 5-8   Producing a List of Duplicate Patient Numbers by Using PROC FREQ

```
PROC FREQ DATA=CLEAN.PATIENTS NOPRINT;
   TABLES PATNO / OUT=DUP_NO(KEEP=PATNO COUNT
                             WHERE=(COUNT GT 1));
RUN;

DATA _NULL_;
   TITLE "Patients with Duplicate Observations";
   FILE PRINT;
   SET DUP_NO;
   PUT "Patient number " PATNO "has " COUNT "observation(s).";
RUN;
```

This program is considerably more efficient than the program requiring sorts and merging. You can also use PROC PRINT instead of the DATA _NULL_ step, if you prefer. The output produced by Program 5-8 is shown next.

```
Patients with Duplicate Observations

Patient number 002 has 2 observation(s).
Patient number 003 has 2 observation(s).
Patient number 006 has 2 observation(s).
```

Selecting Patients with Duplicate Observations by Using a Macro List and SQL

Another quick, easy, and efficient way to select observations with duplicate ID's is to create a macro variable that contains all the patient ID's in the duplicate data set (DUP_NO). Using a short SQL step, you can create a list of patient numbers, separated by spaces or commas (both will work) and placed in quotes, that can subsequently be used as the argument in an IN statement. Program 5-9 demonstrates this.

Program 5-9 Using PROC SQL to Create a List of Duplicates

```
PROC SQL NOPRINT;
   SELECT QUOTE(PATNO)     ❶
      INTO :DUP_LIST SEPARATED BY " "    ❷
      FROM DUP_NO;
QUIT;

PROC PRINT DATA=CLEAN.PATIENTS;
   WHERE PATNO IN (&DUP_LIST);    ❸
   TITLE "Duplicates Selected Using SQL and a Macro Variable";
RUN;
```

The SELECT statement uses the QUOTE function, which places the patient numbers in quotes ❶. Line ❷ assigns the list of patient numbers to a macro variable (DUP_LIST) and separates each of the quoted values with a space. Finally, following a PROC PRINT statement, you supply a WHERE statement ❸ that selects only those patient numbers that are contained in the list of duplicate patient numbers. Notice that this method does not require a sort or a DATA step. Inspection of the output from Program 5-9 shows that the observations are in the original order of the observations in the PATIENTS data set.

```
Duplicates Selected Using SQL and a Macro Variable

Obs    PATNO    GENDER        VISIT    HR    SBP    DBP    DX    AE

  2     002       F       11/13/1998    84    120    78     X    0
  3     003       X       10/21/1998    68    190   100     3    1
  6     006               06/15/1999    72    102    68     6    1
 16     002       F       11/13/1998    84    120    78     X    0
 17     003       M       11/12/1999    58    112    74          0
 31     006       F       07/07/1999    82    148    84     1    0
```

If you want this list in sorted order, you can modify the program as shown in Program 5-10.

Program 5-10 Using PROC SQL to Create a List of Duplicates (in Sorted Order)

```
PROC SQL NOPRINT;
   SELECT QUOTE(PATNO)
      INTO :DUP_LIST SEPARATED BY " "
      FROM DUP_NO;
QUIT;

PROC SORT DATA=CLEAN.PATIENTS OUT=TMP;
   WHERE PATNO IN (&DUP_LIST);
   BY PATNO;
RUN;

PROC PRINT DATA=TMP;
   TITLE "Duplicates Selected Using SQL and a Macro Variable";
RUN;
```

To see an SQL solution, see page 162 in Chapter 8.

Identifying Subjects with "n" Observations Each (DATA Step Approach)

Besides identifying duplicates, you may need to verify that there are "n" observations per subject in a raw data file or in a SAS data set. For example, if each patient in a clinical trial was seen twice, you might want to verify that there are two observations for each patient ID in the file or data set. You can accomplish this task by using a DATA step approach or by using PROC FREQ, the same two methods used earlier to detect duplicates. First, let's look at the DATA step approach.

The key to this approach is the use of the variables FIRST. and LAST., which are created when you use a SET statement followed by a BY statement. To test the programs, let's use the data set PATIENTS2, which was created previously.

Inspection of the PATIENTS2 listing on page 112 shows that patient 002 has three observations, patients 001, 004, 005, and 007 have two observations, and patients 003 and 006 have only one observation.

Program 5-11 lists all the patient ID's who do not have exactly two observations each.

Program 5-11 Using a DATA Step to List All ID's for Patients Who Do Not Have Exactly Two Observations

```
PROC SORT DATA=CLEAN.PATIENTS2(KEEP=PATNO) OUT=TMP;
   BY PATNO;
RUN;

DATA _NULL_;
   TITLE "Patient ID's for Patients with Other than Two Observations";
   FILE PRINT;
   SET TMP;
   BY PATNO;   ❶
   IF FIRST.PATNO THEN N = 1;   ❷
   ELSE N + 1;   ❸
   IF LAST.PATNO AND N NE 2 THEN PUT
      "Patient number " PATNO "has " N "observation(s).";   ❹
RUN;
```

The first step is to sort the data set by patient ID (PATNO) because you will use a BY PATNO statement in the DATA_NULL_ step ❶. In this example, you create a new data set to hold the sorted observations. You may want your original data set sorted by patient ID anyway, in which case you don't have to create a new, temporary data set. Notice also that you only need to keep the patient ID variable in the temporary data set (TMP) because that provides sufficient information to count the number of observations per patient. If each observation contains a large number of variables, this will save processing time.

Use a DATA _NULL_ step to do the counting and output the invalid patient ID's, again as an efficiency technique. When you are processing the first observation for any patient, the temporary variable FIRST.PATNO will be true and N will be set to 1 ❷. Otherwise, the observation counter is incremented ❸. Note that because of line ❸, the variable N is automatically retained, therefore a RETAIN statement is not needed. The fact that N is also initialized to 0 does not matter because you set it to 1 for each new patient ID. Finally, when you reach the last observation for any patient ID ❹, output an error statement if the observation counter is not equal to 2. The next listing shows that this program works properly.

```
Patient ID's for Patients with Other than Two Observations

Patient number 002 has 3 observation(s).
Patient number 003 has 1 observation(s).
Patient number 006 has 1 observation(s).
```

What if a patient with two observations really only had one visit but the single observation was duplicated by mistake? You should probably run one of the programs to detect duplicates before running this program.

Identifying Subjects with "n" Observations Each (Using PROC FREQ)

You can use PROC FREQ to count the number of observations per subject, just as you did to detect duplicates. Use the variable COUNT to determine the number of observations for each value of PATNO, as shown in Program 5-12.

Program 5-12 Using PROC FREQ to List All ID's for Patients Who Do Not Have Exactly Two Observations

```
PROC FREQ DATA=CLEAN.PATIENTS2 NOPRINT;
   TABLES PATNO / OUT=DUP_NO(KEEP=PATNO COUNT
                            WHERE=(COUNT NE 2));
RUN;


DATA _NULL_;
   TITLE "Patient ID's for Patients with Other than Two Observations";
   FILE PRINT;
   SET DUP_NO;
   PUT "Patient number " PATNO "has " COUNT "observation(s).";
RUN;
```

The output data set from PROC FREQ (DUP_NO) contains the variables PATNO and the frequency (COUNT), and there is one observation for each patient who did not have exactly two visits. All that is left to do is to search each observation in the DATA _NULL_ step and print an error message. Output from Program 5-12 is shown next.

```
Patient ID's for Patients with Other than Two Observations

Patient number 002 has 3 observation(s).
Patient number 003 has 1 observation(s).
Patient number 006 has 1 observation(s).
```

It is usually easier to let a PROC do the work, as in this example, rather than doing all the work yourself with a DATA step. See Program 8-10, on page 163, for an SQL solution to this program.

6 **Working with Multiple Files**

Introduction

This chapter addresses data validation techniques where multiple files or data sets are involved. You will see how to verify that a given ID exists in multiple files and how to verify some more complicated multi-file rules. Because it is impossible to anticipate all the possible multi-file rules you may need, much of this chapter is devoted to programming tips and techniques rather than actual examples that you can use directly.

Checking for an ID in Each of Two Files

One requirement of a large project may be that a particular ID value exists in each of several SAS data sets. Let's start out by demonstrating how you can easily check that an ID is in each of two files. This will be generalized later to include an arbitrary number of files.

The technique demonstrated in this section is to merge the two files in question, using the ID variable as a BY variable. The key to the program is the IN= data set option that sets a logical variable to true or false, depending on whether or not the data set provides values to the current observation being created. An example will make this clearer. In Program 6-1 are the SAS statements to create two SAS data sets for testing purposes.

Program 6-1   Creating Two Test Data Sets for Chapter 6 Examples

```
DATA ONE;
   INPUT PATNO X Y;
DATALINES;
1 69 79
2 56 .
3 66 99
5 98 87
12 13 14
;
DATA TWO;
   INPUT PATNO Z;
DATALINES;
1 56
3 67
4 88
5 98
13 99
;
```

Notice that ID's 2 and 12 are in data set ONE but not in data set TWO; ID's 4 and 13 are in data set TWO but not in data set ONE. Program 6-2 gives detailed information on the unmatched ID's.

Program 6-2   Identifying ID's Not in Each of Two Data Sets

```
PROC SORT DATA=ONE;     ❶
   BY PATNO;
RUN;

PROC SORT DATA=TWO;     ❷
   BY PATNO;
RUN;

DATA _NULL_;
   FILE PRINT;
   TITLE "Listing of Missing ID's";
   MERGE ONE(IN=INONE)
         TWO(IN=INTWO)   END=LAST;   ❸
   BY PATNO;   ❹
```

```
IF NOT INONE THEN DO;   ❺
   PUT "ID " PATNO "is not in data set ONE";
   N + 1;
END;

ELSE IF NOT INTWO THEN DO;   ❻
   PUT "ID " PATNO "is not in data set TWO";
   N + 1;
END;

IF LAST AND N EQ 0 THEN
   PUT "All ID's Match in Both Files";   ❼
RUN;
```

Before you can merge the two data sets, they must first be sorted by the BY variable ❶, ❷. The MERGE statement ❸ is the key to this program. Each of the data set names is followed by the data set option IN=*logical_variable*. In addition, the END= *variable_name* option creates a logical variable that is set to 1 (true) when the last observation from all data sets has been processed. Using a MERGE statement would be useless in this application without the BY statement ❹. One final note: although this program runs correctly even if there are multiple observations with the same BY variable in both data sets, you would be wise to check for unexpected duplicates as described in the previous chapter or to use the NODUPKEY option of PROC SORT on one of the data sets.

Let's "play computer" to see how this program works. Both data sets contain an observation for PATNO = 1. Therefore, the two logical variables INONE and INTWO are both true and neither of the IF statements ❺,❻ is true and a message is not printed to the output file. The next value of PATNO is 2, from data set ONE. Because this value is not in data set TWO, the values of INONE and INTWO are true and false, respectively. Therefore, statement ❻ is true and the appropriate message is printed to the output file. When you reach PATNO = 4, which exists in data set TWO but not in data set ONE, statement ❺ is true and its associated message is printed out. Note that anytime a value of PATNO is missing from one of the files, the variable N is incremented. If you reach the last observation in the files being merged, the logical variable LAST is true. If there are no ID errors, N will still be 0 and statement ❼ will be true. When you run Program 6-2, the following output is obtained.

```
Listing of Missing ID's

ID 2 is not in data set TWO
ID 4 is not in data set ONE
ID 12 is not in data set TWO
ID 13 is not in data set ONE
```

All the ID errors are correctly displayed. If you want to extend this program to more than two data sets, the program could become long and tedious. A macro approach can be used to accomplish the ID checking task with an arbitrary number of data sets. The next section demonstrates such a program.

Checking for an ID in Each of "n" Files

Data set THREE is added to the mix to demonstrate how to approach this problem when there are more than two data sets. First, run Program 6-3 to create the new data set THREE.

Program 6-3   Creating a Third Data Set for Testing Purposes

```
DATA THREE;
   INPUT PATNO GENDER $ @@;
DATALINES;
1 M 2 F 3 M 5 F 6 M 12 M 13 M
;
```

Before developing a macro, let's look at Program 6-4, which is a rather simple but slightly tedious program to accomplish the ID checks.

Program 6-4   Checking for an ID in Each of Three Data Sets (Long Way)

```
PROC SORT DATA=ONE(KEEP=PATNO) OUT=TMP1;
   BY PATNO;
RUN;

PROC SORT DATA=TWO(KEEP=PATNO) OUT=TMP2;
   BY PATNO;
RUN;

PROC SORT DATA=THREE(KEEP=PATNO) OUT=TMP3;
   BY PATNO;
RUN;
```

```
DATA _NULL_;
   FILE PRINT;
   TITLE "Listing of Missing ID's and Data Set Names";
   MERGE TMP1(IN=IN1)
         TMP2(IN=IN2)
         TMP3(IN=IN3)  END=LAST;
   BY PATNO;

   IF NOT IN1 THEN DO;
      PUT "ID " PATNO "missing from data set ONE";
      N + 1;
   END;

   IF NOT IN2 THEN DO;
      PUT "ID " PATNO "missing from data set TWO";
      N + 1;
   END;

   IF NOT IN3 THEN DO;
      PUT "ID " PATNO "missing from data set THREE";
      N + 1;
   END;

   IF LAST AND N EQ 0 THEN
      PUT "All ID's Match in All Files";

RUN;
```

Program 6-4 can be extended to accommodate any number of data sets. The output is shown next.

```
Listing of Missing ID's and Data Set Names

ID 2 missing from data set TWO
ID 4 missing from data set ONE
ID 4 missing from data set THREE
ID 6 missing from data set ONE
ID 6 missing from data set TWO
ID 12 missing from data set TWO
ID 13 missing from data set ONE
```

A pattern begins to emerge. Notice the sorts and IF statements follow a pattern that can be automated by writing a macro. Because the programming can become complicated, two macros will be written to accomplish this task. The first is easier to understand but not as elegant — the second is more complicated but slightly more flexible and easier to run.

A Simple Macro to Check ID's in Multiple Files

The first macro has as its calling arguments the ID variable name and the data set names. In this program, up to 10 data sets are allowed. This can easily be increased if necessary. The more complicated macro, developed in Program 6-6, does not have this limitation. Program 6-5 is the simple macro program. An explanation follows.

Program 6-5   Creating a Macro to Check for an ID in Each of "n" Files (Simple Way)

```
*-------------------------------------------------------------*
| Program Name: ID_SIMP.SAS  in C:\CLEANING                   |
| Purpose: Simple version of the macro to test if an ID exists in |
|          each of up to 10 data sets                         |
| Arguments: ID     - Name of the ID variable                 |
|            DSNn   - Name of the nth data set                |
| Example:  %ID_SIMP(PATNO,ONE,TWO,THREE)                     |
*-------------------------------------------------------------*;
%MACRO ID_SIMP(ID,DSN1,DSN2,DSN3,DSN4,DSN5,
               DSN6,DSN7,DSN8,DSN9,DSN10);
   TITLE "Report of ID's Not in All Data Sets";
   ***Sorting section;
   %DO I = 1 %TO 10;   ❶
      %IF &&DSN&I NE %THEN %DO; /* If non-null argument */   ❷
         %LET N_DATA = &I;   ❸
         PROC SORT DATA = &&DSN&I(KEEP=&ID) OUT=TMP&I;   ❹
            BY &ID;
         RUN;
      %END;
      %ELSE %LET I = 10;   /* Stop the loop when DSNn is null */   ❺
   %END;

   ***Create MERGE statements;
   DATA _NULL_;
      FILE PRINT;
      MERGE

      %DO I = 1 %TO &N_DATA;   ❻
         &&DSN&I(IN=IN&I)
      %END;

      END=LAST;   ❼
      ***End MERGE statement;
```

```
    BY &ID;

    ***Error reporting section;
    %DO I = 1 %TO &N_DATA;  ❽
       IF NOT IN&I THEN DO;
          PUT "ID " &ID "Missing from data set &&DSN&I";
          N + 1;
       END;
    %END;

    IF LAST AND N EQ 0 THEN DO;   ❾
       PUT "All ID's Match in All Files";
       STOP;
    END;

  RUN;

%MEND ID_SIMP;
```

To run this macro for three data sets (ONE, TWO, and THREE) with a common ID variable (PATNO), you would write

```
%ID_SIMP(PATNO,ONE,TWO,THREE)
```

The program starts with a MACRO statement with calling arguments for the name of the common ID variable and up to 10 data set names. Next, you want to sort all the data sets by the ID variable. The %DO loop ❶ accomplishes this task. When I is equal to 1, the macro variable &&DSN&I ❹ first resolves to &DSN1, which is the macro variable representing the first data set to be sorted. The same task is performed for each of the "n" data sets. Each time the loop iterates, the value of N_DATA is set equal to the loop index ❸. When all the data sets have been processed, the value of &&DSN&I will be null and the %IF statement ❷ will be false. This causes the loop counter to be set to 10, which stops the loop ❺.

The next section of the program generates statements similar to those following the sort in Program 6-4. After the word "MERGE," a %DO loop ❻ writes the data set names followed by a matching IN= option. For example, if the first data set is called ONE, the first iteration of the %DO loop writes the text "ONE(IN=IN1)." After the last data set name and IN= option, the text "END=LAST;" finishes the MERGE statement ❼.

If any of the IN= logical variables are false (that data set did not have a contribution to the MERGE), an error report is generated ❽. Finally, if all the data sets have been merged (LAST is true) and N still equals 0 (no missing ID's), the IF statement in line ❾ is true and the program reports that all ID's match.

To help clarify this program, look at the code generated by the macro (using the MPRINT system option) when it is used to check the three data sets ONE, TWO, and THREE.

SAS Log Showing the Macro Generate Program (Using the MPRINT Option)

```
MPRINT(ID_SIMP):    TITLE "Report of ID's Not in All Data Sets";
MPRINT(ID_SIMP):    ***Sorting section;
MPRINT(ID_SIMP):    PROC SORT DATA = one(KEEP=patno) OUT=TMP1;
MPRINT(ID_SIMP):    BY patno;
MPRINT(ID_SIMP):    RUN;

MPRINT(ID_SIMP):    PROC SORT DATA = two(KEEP=patno) OUT=TMP2;
MPRINT(ID_SIMP):    BY patno;
MPRINT(ID_SIMP):    RUN;

MPRINT(ID_SIMP):    PROC SORT DATA = three(KEEP=patno) OUT=TMP3;
MPRINT(ID_SIMP):    BY patno;
MPRINT(ID_SIMP):    RUN;

MPRINT(ID_SIMP):    ***Create MERGE statements;
MPRINT(ID_SIMP):    DATA _NULL_;
MPRINT(ID_SIMP):    FILE PRINT;
MPRINT(ID_SIMP):    MERGE one(IN=IN1) two(IN=IN2) three(IN=IN3) END=LAST;
MPRINT(ID_SIMP):    ***End MERGE statement;
MPRINT(ID_SIMP):    BY patno;
MPRINT(ID_SIMP):    ***Error reporting section;
MPRINT(ID_SIMP):    IF NOT IN1 THEN DO;
MPRINT(ID_SIMP):    PUT "ID " patno "Missing from data set one";
MPRINT(ID_SIMP):    N + 1;
MPRINT(ID_SIMP):    END;
MPRINT(ID_SIMP):    IF NOT IN2 THEN DO;
MPRINT(ID_SIMP):    PUT "ID " patno "Missing from data set two";
MPRINT(ID_SIMP):    N + 1;
MPRINT(ID_SIMP):    END;
MPRINT(ID_SIMP):    IF NOT IN3 THEN DO;
MPRINT(ID_SIMP):    PUT "ID " patno "Missing from data set three";
```

```
MPRINT(ID_SIMP):    N + 1;
MPRINT(ID_SIMP):    END;
MPRINT(ID_SIMP):    IF LAST AND N EQ 0 THEN DO;
MPRINT(ID_SIMP):    PUT "All ID's Match in All Files";
MPRINT(ID_SIMP):    STOP;
MPRINT(ID_SIMP):    END;
MPRINT(ID_SIMP):    RUN;
```

Notice that the macro generated code is basically the same as the non-macro program that preceded it. Finally, here is the output from Program 6-5.

```
Report of ID's Not in All Data Sets

ID 2 Missing from data set two
ID 4 Missing from data set one
ID 4 Missing from data set three
ID 6 Missing from data set one'
ID 12 Missing from data set two
ID 13 Missing from data set one
```

A More Complicated Multi-File Macro for ID Checking

The above macro can be modified to handle any number of data sets. This is accomplished by calling the macro with one argument representing the ID variable and the other, a list of data sets to be checked, separated by spaces. You can then use the %SCAN macro function to extract the individual data set names from the list of data set names and create a macro variable for each of the data set names. Program 6-6 presents the macro. An explanation follows.

Program 6-6   Writing a More General Macro to Handle Any Number of Data Sets

```
*-----------------------------------------------------------------*
| Program Name: CHECK_ID.SAS  in C:\CLEANING                      |
| Purpose: Macro which checks if an ID exists in each of n files  |
| Arguments: The name of the ID variable, followed by as many     |
|            data set names as you want, separated by spaces      |
| Example: %CHECK_ID(PATNO,ONE TWO THREE)                         |
*-----------------------------------------------------------------*;
```

```
%MACRO CHECK_ID(ID,DS_LIST);   ❶
   %LET STOP = 999;  /* Initialize stop at a large value */
   %DO I = 1 %TO &STOP;
      %LET DSN = %SCAN(&DS_LIST,&I);  /* Break up list
                                         into data set names */
      %IF &DSN NE %THEN  /* If non-null argument */  ❷
      %DO;
         %LET N = &I;    /* the number of data sets */
         PROC SORT DATA=&DSN(KEEP=&ID) OUT=TMP&I;
            BY &ID;
         RUN;
      %END;
      %ELSE %LET I = &STOP;  /* Set index to max so loop stops */
   %END;

   DATA _NULL_;
      FILE PRINT;
      MERGE

      %DO I = 1 %TO &N;
         TMP&I(IN=IN&I)
      %END;

      END=LAST;
      BY &ID;

      %DO I = 1 %TO &N;
         %LET DSN = %SCAN(&DS_LIST,&I);
         IF NOT IN&I THEN DO;
            PUT "ID " &ID "missing from data set &DSN";
            NN + 1;
         END;
      %END;

      IF LAST AND NN EQ 0 THEN DO;
         PUT "All ID's Match in All Files";
         STOP;
      END;

      RUN;

%MEND CHECK_ID;
```

This macro is similar to the previous one except for the way the multiple data set names are handled. The second argument in the macro call is a list of data set names, separated by spaces ❶. The %SCAN function, which interprets spaces as delimiters (the default), allows you to extract the name of the multiple data set names. When the %SCAN function returns a null string ❷, the loop ends (by setting the value of the DO loop counter to the ending value) and the program continues. The remainder of the program is similar to the previous program where a MERGE statement is constructed, using the multiple data set names followed by the IN= data set option.

To invoke this macro to check the three data sets (ONE, TWO, and THREE) with the ID variable PATNO, you would write:

```
%CHECK_ID(PATNO,ONE TWO THREE)
```

Remember that the multiple data set names are separated by spaces rather than commas as in the previous macro.

More Complicated Multi-File Rules

Every data collection project will have its own set of rules. The programs in this section are intended to demonstrate techniques rather than be exact models for multiple file data validation rules.

In this first example, you want to be sure an observation has been added to the laboratory test data set (LAB_TEST) if there was an adverse event of 'X' entered for any patient ID in the adverse event data set (AE). The AE and the LAB_TEST data sets can be created by running the programs given in the Appendix.

Here are the listings of the AE and LAB_TEST data sets.

```
Listing of Data Set AE

PATNO        DATE_AE     A_EVENT

 001       11/21/1998       W
 001       12/13/1998       Y
 003       11/18/1998       X
 004       09/18/1998       O
 004       09/19/1998       P
 011       10/10/1998       X
 013       09/25/1998       W
 009       12/25/1998       X
 022       10/01/1998       W
 025       02/09/1999       X
```

Note that each patient ID (PATNO) may have more than one adverse event.

```
Listing of data set LAB_TEST

PATNO        LAB_DATE     WBC        RBC

 001       11/15/1998     9000      5.45
 003       11/19/1998     9500      5.44
 007       10/21/1998     8200      5.23
 004       12/22/1998    11000      5.55
 025       01/01/1999     8234      5.02
 022       10/10/1998     8000      5.00
```

Note that each patient has only one observation in this data set.

According to our rule, any patient with an adverse event of 'X' should have one observation in the LAB_TEST data set. Patients 003, 011, 009, and 025 all had an adverse event of 'X', however, only patients 003 and 025 had an observation in the LAB_TEST data set (although the date of the lab test for patient 025 is earlier than the date of the AE, let's ignore this for now). One approach to locating these two omissions is to merge the AE and LAB_TEST data sets by patient ID, selecting only those patients with an AE of 'X'. Then, using the IN= data set option on the merge, you can locate any missing observations. This is shown in Program 6-7. An explanation follows.

**Program 6-7   Verifying That Patients with an Adverse Event of "X" in Data Set
AE have an Entry in Data Set LAB_TEST**

```
PROC SORT DATA=CLEAN.AE OUT=AE_X;
   WHERE A_EVENT = 'X';    ❶
   BY PATNO;
RUN;

PROC SORT DATA=CLEAN.LAB_TEST(KEEP=PATNO LAB_DATE) OUT=LAB;
   BY PATNO;
RUN;

DATA MISSING;
   MERGE AE_X
         LAB(IN=IN_LAB);    ❷
   BY PATNO;
   IF NOT IN_LAB;
RUN;

PROC PRINT DATA=MISSING LABEL;
   TITLE "Patients with AE of X Who Are Missing Lab Test Entry";
   ID PATNO;
   VAR DATE_AE A_EVENT;
RUN;
```

Each of the two data sets is first sorted by patient ID (PATNO). In addition, by using a
WHERE statement ❶ following the PROC SORT statement, only those observations in
the adverse events data set with event 'X' are selected. The key to the program is
statement ❷, where you use the IN= option to create the temporary logical variable
IN_LAB. Because there should be an observation in LAB_TEST for every patient with
an adverse event of 'X', anytime the logical variable IN_LAB is false, you have located a
patient with a missing laboratory test. The output from Program 6-7 is shown next.

```
Patients with AE of X Who Are Missing Lab Test Entry

Patient                 Adverse
  ID     Date of AE     Event

  009    12/25/1998       X
  011    10/10/1998       X
```

Checking That the Dates Are in the Proper Order

It was mentioned earlier that the date of the laboratory test for patient 025 was prior to the date of the adverse event. You should modify your program to detect this. Remembering that dates are just numbers (the number of days from January 1, 1960), you can easily compare the two dates in the last DATA step, as shown in Program 6-8.

Program 6-8   Adding the Condition That the Lab Test Must Follow the Adverse Event

```
TITLE  "Patients with AE of X Who Are Missing Lab Test Entry";
TITLE2 "or the Date of the Lab Test Is Earlier Than the AE Date";
TITLE3 "-------------------------------------------------------";

DATA _NULL_;
   FILE PRINT;
   MERGE AE_X(IN=IN_AE)
         LAB(IN=IN_LAB);

   BY PATNO;
   IF NOT IN_LAB THEN PUT
      "No Lab Test for Patient " PATNO "with Adverse Event X";
   ELSE IF IN_AE AND LAB_DATE EQ . THEN PUT  ❶
      "Date of Lab Test Is Missing for Patient "
      PATNO /
      "Date of AE Is " DATE_AE /;
   ELSE IF IN_AE AND LAB_DATE LT DATE_AE THEN PUT  ❷
      "Date of Lab Test Is Earlier Than Date of AE for Patient "
      PATNO /
      "  Date of AE Is " DATE_AE " Date of Lab Test Is " LAB_DATE /;
RUN;
```

One IF statement ❶ checks if the laboratory date is missing, and the other IF statement ❷ tests if the laboratory date is prior to (less than) the date of the adverse event and an appropriate message is printed. Output from this program is shown next.

```
Patients with AE of X Who Are Missing Lab Test Entry
or the Date of the Lab Test Is Earlier Than the AE Date
--------------------------------------------------------
No Lab Test for Patient 009 with Adverse Event X
No Lab Test for Patient 011 with Adverse Event X
Date of Lab Test Is Earlier Than Date of AE for Patient 025
Date of AE Is 02/09/1999  Date of Lab Test Is 01/01/1999
```

Programs to verify multi-file rules can become very complicated. However, many of the techniques discussed in this chapter should prove useful.

7 Double Entry and Verification (PROC COMPARE)

Introduction

Many critical data applications require that you have the data entered twice and then compare the resulting files for discrepancies. This is usually referred to as double entry and verification. In the "old days," when I was first learning to use computers, most data entry was done using a keypunch (although my boys will tell you that, in my day, it was done with a hammer and chisel on stone tablets). The most common method of double entry and verification was done on a special keypunch machine called a verifier. The original cards were placed in the input hopper and a keypunch operator (preferably not the one who entered the data originally) re-keyed the information from the data entry form. If the information being typed matched the information already punched on the card, the card was accepted and a punch was placed, usually in column 81 of the card. If the information did not match, a check could be made to see whether the error was on the original card or in the re-keying of the information.

Today, there are several programs that accomplish the same goal by having all the data entered twice and then comparing the resulting data files. Some of these programs are quite sophisticated and also quite expensive. SAS software has a very flexible procedure called PROC COMPARE, which can be used to compare the contents of two SAS data sets. You can refer to the *SAS Procedures Guide* in hard copy or online documentation for more information. This chapter presents some simple examples using PROC COMPARE.

Conducting a Simple Comparison of Two Data Sets without an ID Variable

The simplest application of PROC COMPARE is presented first, determining if the contents of two SAS data sets are identical. Suppose you have two people enter data from some coding forms and the two data sets are called FILE_1 and FILE_2. A listing of the two files is shown next.

```
FILE_1

001M10211946130 80
002F12201950110 70
003M09141956140 90
004F10101960180100
007m10321940184110


FILE_2

001M1021194613080
002F12201950110 70
003M09141956144 90
004F10101960180100
007M10231940184110
```

Here is the file format.

Variable	Description	Starting Column	Length	Type
PATNO	Patient Number	1	3	Numeric
GENDER	Gender	4	1	Character
DOB	Date of Birth	5	8	mmddyyyy
SBP	Systolic Blood Pressure	13	3	Numeric
DBP	Diastolic Blood Pressure	16	3	Numeric

The data, without mistakes, should have been:

```
Correct Data Representation

001M10211946130 80
002F12201950110 70
003M09141956140 90
004F10101960180100
007M10231940184110
```

A visual inspection of the two original files shows the following discrepancies.

For patient 001, there is a space missing before the 80 at the end of the line in FILE_2.

For patient 003, SBP is 144 instead of 140 in FILE_2.

For patient 007, the gender is entered in lowercase and the digits are interchanged in the day field of the date in FILE_1.

Let's see how to use PROC COMPARE to detect these differences. You have some choices here. One way to proceed is to create two SAS data sets as shown in Program 7-1.

Program 7-1   Creating Data Sets ONE and TWO from Two Raw Data Files

```
DATA ONE;
   INFILE "C:\CLEANING\FILE_1" PAD;
   INPUT @1   PATNO   3.
         @4   GENDER  $1.
         @5   DOB     MMDDYY8.
         @13  SBP     3.
         @16  DBP     3.;
   FORMAT DOB MMDDYY10.;
RUN;

DATA TWO;
   INFILE "C:\CLEANING\FILE_2" PAD;
   INPUT @1   PATNO   3.
         @4   GENDER  $1.
         @5   DOB     MMDDYY8.
         @13  SBP     3.
         @16  DBP     3.;
   FORMAT DOB MMDDYY10.;
RUN;
```

Then run PROC COMPARE as shown in Program 7-2.

Program 7-2   Running PROC COMPARE

```
PROC COMPARE BASE=ONE COMPARE=TWO;
   TITLE "Using PROC COMPARE to Compare Two Data Sets";
RUN;
```

The procedure options BASE= and COMPARE= identify the two data sets to be compared. In this example, data set ONE was arbitrarily chosen as the base data set. (The option DATA= may be used in place of BASE= because they are equivalent.)

Here is the output from PROC COMPARE .

```
Using PROC COMPARE to Compare Two Data Sets

COMPARE Procedure
Comparison of WORK.ONE with WORK.TWO
(Method=EXACT)

Data Set Summary

Data set            Created           Modified      Nvar    NObs

WORK.ONE   17AUG98:10:30:25   17AUG98:10:30:25        5       5
WORK.TWO   17AUG98:10:30:25   17AUG98:10:30:25        5       5

Variables Summary

Number of Variables in Common: 5.

Observation Summary

Observation       Base   Compare

First  Obs           1         1
First  Unequal       3         3
Last   Unequal       5         5
Last   Obs           5         5


                                                  Continued
```

```
Number of Observations in Common: 5.
Total Number of Observations Read from WORK.ONE: 5.
Total Number of Observations Read from WORK.TWO: 5.

Number of Observations with Some Compared Variables Unequal: 2.
Number of Observations with All Compared Variables Equal: 3.

Values Comparison Summary

Number of Variables Compared with All Observations Equal: 2.
Number of Variables Compared with Some Observations Unequal: 3.
Number of Variables with Missing Value Differences: 1.
Total Number of Values which Compare Unequal: 3.
Maximum Difference: 4.

Using PROC COMPARE to Compare Two Data Sets

COMPARE Procedure
Comparison of WORK.ONE with WORK.TWO
(Method=EXACT)

Variables with Unequal Values

Variable   Type   Len   Ndif    MaxDif   MissDif

GENDER     CHAR    1     1                  0
DOB        NUM     8     1        0         1
SBP        NUM     8     1      4.000       0

Value Comparison Results for Variables
```

Obs	Base Value GENDER	Compare Value GENDER
	—	—
5	m	M

Obs	Base DOB	Compare DOB	Diff.	% Diff
5	.	10/23/40	.	.

Obs	Base SBP	Compare SBP	Diff.	% Diff
3	140.0000	144.0000	4.0000	2.8571

Notice that the left-adjusted value of 80 for patient 001 in FILE_1 was not flagged as an error. Why? Because SAS correctly reads left-adjusted numeric values and the comparison is between the two SAS data sets, not the raw files themselves. Also, the incorrect date of 10/32/1940 (patient number 005 in FILE_1) was shown as a missing value in the output. If you inspect the SAS Log, you will see the incorrect date was flagged as an error. When invalid dates are encountered, SAS substitutes a missing value for the date. If you do not want this to happen, you can use a character informat instead of a date informat for data checking purposes.

You may find that the output from PROC COMPARE contains more information than you need. The procedure option BRIEF reduces the output considerably by producing only a short comparison summary and suppressing the lengthy output shown after Program 7-2. Here is the output from PROC COMPARE using the BRIEF option.

```
Using PROC COMPARE to Compare Two Data Sets

COMPARE Procedure
Comparison of WORK.ONE with WORK.TWO
(Method=EXACT)

NOTE: Values of the following 3 variables compare unequal: GENDER
      DOB SBP

Value Comparison Results for Variables

            ||  Base Value        Compare Value
    Obs     ||  GENDER            GENDER
  _____  ||  ___               ___
            ||
      5     ||  m                 M



            ||  Base      Compare
    Obs     ||  DOB       DOB          Diff.       % Diff
  _____  ||  _____   _____    _____     _____
            ||
      5     ||     .      10/23/40       .            .



            ||  Base      Compare
    Obs     ||  SBP       SBP          Diff.       % Diff
  _____  ||  _____   _____    _____     _____
            ||
      3     ||  140.0000  144.0000    4.0000       2.8571
```

If you want to simply emulate a data-entry, verify process, you can proceed in another way. You can treat each line of each file as a long character string as shown in Program 7-3.

Program 7-3   Using PROC COMPARE to Compare Two Data Records

```
DATA ONE;
   INFILE "C:\CLEANING\FILE_1" PAD;
   INPUT STRING $CHAR18.;
RUN;

DATA TWO;
   INFILE "C:\CLEANING\FILE_2" PAD;
   INPUT STRING $CHAR18.;
RUN;

PROC COMPARE BASE=ONE COMPARE=TWO BRIEF;
   TITLE "Treating Each Line as a String";
RUN;
```

This greatly simplifies the DATA steps and, perhaps, gives you a result closer to what you really want, an exact comparison of the raw data files. Here is the output.

```
Treating Each Line as a String

COMPARE Procedure
Comparison of WORK.ONE with WORK.TWO
(Method=EXACT)

NOTE: Values of the following 1 variables compare unequal: STRING

Value Comparison Results for Variables

                Base Value            Compare Value
      Obs       STRING                STRING
     _____   _____     _____

         1      001M10211946130 80    001M1021194613080
         3      003M09141956140 90    003M09141956144 90
         5      007m10321940184110    007M10231940184110
```

Of course, you now have to look over the two lines and determine where the differences are. Which of these methods you use will depend on your goals in the verification process.

Using PROC COMPARE with an ID Variable

If you are going to use the first method described in Programs 7-1 and 7-2, identifying each of the variables in the two data sets, it is much better to specify an ID variable (PATNO, in this example) to link the two files. You specify the ID variable with an ID statement. Without an ID variable, PROC COMPARE does an observation-by-observation match between the two files. If there are different numbers of observations in the two files, this method is quite useless. Let's rerun PROC COMPARE on the two data sets and add the ID variable PATNO as shown in Program 7-4.

Program 7-4   Using PROC COMPARE with an ID Variable

```
PROC COMPARE BASE=ONE COMPARE=TWO;
   TITLE "Using PROC COMPARE to Compare Two Data Sets";
   ID PATNO;
RUN;
```

```
Using PROC COMPARE to Compare Two Data Sets

COMPARE Procedure
Comparison of WORK.ONE with WORK.TWO
(Method=EXACT)
Data Set Summary

Dataset              Created           Modified    NVar    NObs

WORK.ONE   13AUG98:10:49:13   13AUG98:10:49:13       5       5
WORK.TWO   13AUG98:10:49:13   13AUG98:10:49:13       5       5


Variables Summary

Number of Variables in Common: 5.
Number of ID Variables: 1.


Observation Summary

Observation      Base   Compare   ID

First  Obs          1         1   PATNO=1
First  Unequal      3         3   PATNO=3
Last   Unequal      5         5   PATNO=7
Last   Obs          5         5   PATNO=7
                                               Continued
```

```
Number of Observations in Common: 5.

Total Number of Observations Read from WORK.ONE: 5.
Total Number of Observations Read from WORK.TWO: 5.

Number of Observations with Some Compared Variables Unequal: 2.
Number of Observations with All Compared Variables Equal: 3.


Values Comparison Summary

Number of Variables Compared with All Observations Equal: 1.
Number of Variables Compared with Some Observations Unequal: 3.
Number of Variables with Missing Value Differences: 1.
Total Number of Values which Compare Unequal: 3.
Maximum Difference: 4.

Using PROC COMPARE to Compare Two Data Sets

COMPARE Procedure
Comparison of WORK.ONE with WORK.TWO
(Method=EXACT)

Variables with Unequal Values

Variable   Type   Len   Ndif    MaxDif   MissDif

GENDER     CHAR    1     1                  0
DOB        NUM     8     1         0        1
SBP        NUM     8     1      4.000       0


Value Comparison Results for Variables
```

PATNO	Base Value GENDER	Compare Value GENDER
	—	—
7	m	M

PATNO	Base DOB	Compare DOB	Diff.	% Diff
7	.	10/23/40	.	.

PATNO	Base SBP	Compare SBP	Diff.	% Diff
3	140.0000	144.0000	4.0000	2.8571

This output has the advantage of identifying the incompatible data lines by patient number, making it easier to go back to the data sheets and determining the correct values.

Using PROC COMPARE with Two Data Sets That Have an Unequal Number of Observations

The ID statement is especially useful when the two data sets do not contain the same number of observations or when there is a discrepancy between the values of the ID variables. To see how PROC COMPARE treats this problem, look at the two new files (FILE_1B and FILE_2B) shown next. A new patient number (005) has been added to FILE_1 to make FILE_1B, and patient number 004 has been omitted from FILE_2 to make FILE_2B.

```
FILE_1B

001M10211946130 80
002F12201950110 70
003M09141956140 90
004F10101960180100
005M01041930166 88
007m10321940184110

FILE_2B

001M1021194613080
002F12201950110 70
003M09141956144 90
007M10231940184110
```

The two SAS data sets (ONE_B and TWO_B) are created as shown in Program 7-1. You can then run PROC COMPARE, using PATNO as the ID variable, as shown in Program 7-5.

Program 7-5   Running PROC COMPARE on Two Data Sets of Different Length

```
PROC COMPARE BASE=ONE_B COMPARE=TWO_B;
   TITLE "Comparing Two Data Sets with Different ID Values";
   ID PATNO;
RUN;
```

Here is the output from Program 7-5.

```
Comparing Two Data Sets with Different ID Values

COMPARE Procedure
Comparison of WORK.ONE_B with WORK.TWO_B
(Method=EXACT)

Data Set Summary

Data set                Created             Modified    NVar    NObs

WORK.ONE_B   13AUG98:11:12:13   13AUG98:11:12:13     5       6
WORK.TWO_B   13AUG98:11:12:14   13AUG98:11:12:14     5       4


Variables Summary

Number of Variables in Common: 5.
Number of ID Variables: 1.


Observation Summary

Observation      Base   Compare   ID

First  Obs          1         1   PATNO=1
First  Unequal      3         3   PATNO=3
Last   Unequal      6         4   PATNO=7
Last   Obs          6         4   PATNO=7

Number of Observations in Common: 4.
Number of Observations in WORK.ONE_B but not in WORK.TWO_B: 2.
Total Number of Observations Read from WORK.ONE_B: 6.
Total Number of Observations Read from WORK.TWO_B: 4.

Number of Observations with Some Compared Variables Unequal: 2.
Number of Observations with All Compared Variables Equal: 2.

Values Comparison Summary

Number of Variables Compared with All Observations Equal: 1.
Number of Variables Compared with Some Observations Unequal: 3.
Number of Variables with Missing Value Differences: 1.
Total Number of Values That Compare Unequal: 3.
Maximum Difference: 4.
                                                   Continued
```

```
Comparing Two Data Sets with Different ID Values

COMPARE Procedure
Comparison of WORK.ONE_B with WORK.TWO_B
(Method=EXACT)

Variables with Unequal Values

Variable   Type   Len   Ndif    MaxDif   MissDif

GENDER     CHAR    1     1                  0
DOB        NUM     8     1        0         1
SBP        NUM     8     1      4.000       0



Value Comparison Results for Variables
```

PATNO	Base Value GENDER	Compare Value GENDER
	—	—
7	m	M

PATNO	Base DOB	Compare DOB	Diff.	% Diff
7	.	10/23/40	.	.

PATNO	Base SBP	Compare SBP	Diff.	% Diff
3	140.0000	144.0000	4.0000	2.8571

Notice that the information concerning the missing patients is not shown. To see this, add the two procedure options LISTBASE and LISTCOMP to see a list of observations found in one data set but not in the other. Next, you'll see the additional information that you will get when these two options are used.

```
Comparing Two Data Sets with Different ID Values

COMPARE Procedure
Comparison of WORK.ONE_B with WORK.TWO_B
(Method=EXACT)

Comparison Results for Observations

Observation 4 in WORK.ONE_B not found in WORK.TWO_B: PATNO=4.

Observation 5 in WORK.ONE_B not found in WORK.TWO_B: PATNO=5.
```

Comparing Two Data Sets When Some Variables Are Not in Both Data Sets

PROC COMPARE can also be used to compare selected variables between two data sets. Suppose you have one data set that contains demographic information on each patient in a clinical trial (DEMOG). In addition, you have another file from a previous study that contains some of the same patients and some of the same demographic information (OLDDEMOG). Program 7-6 creates these sample data sets.

Program 7-6   Creating Two Test Data Sets, DEMOG and OLDDEMOG

```
***Program to create data sets DEMOG and OLDDEMOG;
DATA DEMOG;
   INPUT  @1  PATNO  3.
          @4  GENDER $1.
          @5  DOB    MMDDYY10.
          @15 HEIGHT 2.;
   FORMAT DOB MMDDYY10.;
DATALINES;
001M10/21/194668
003F11/11/105062
004M04/05/193072
006F05/13/196863
;
```

```
DATA OLDDEMOG;
   INPUT @1  PATNO   3.
         @4  DOB     MMDDYY8.
         @12 GENDER  $1.
         @13 WEIGHT  3.;
   FORMAT DOB MMDDYY10.;
DATALINES;
00110211946M155
00201011950F102
00404051930F101
00511111945M200
00605131966F133
;
```

You want to compare the date of birth (DOB) and gender (GENDER) for each patient. PROC COMPARE will automatically compare all variables that are in common to both data sets, so you can proceed as shown in Program 7-7.

Program 7-7   Comparing Two Data Sets That Contain Different Variables

```
PROC COMPARE BASE=OLDDEMOG COMPARE=DEMOG BRIEF;
   TITLE "Comparing Demographic Information between Two Data Sets";
   ID PATNO;
RUN;
```

Here is the output from this procedure.

```
Comparing Demographic Information between Two Data Sets

COMPARE Procedure
Comparison of WORK.OLDDEMOG with WORK.DEMOG
(Method=EXACT)

NOTE: Data set WORK.OLDDEMOG contains 2 observations not in
      WORK.DEMOG.
NOTE: Data set WORK.DEMOG contains 1 observations not in
      WORK.OLDDEMOG.
NOTE: Values of the following 2 variables compare unequal: DOB
      GENDER

                                                    Continued
```

```
Value Comparison Results for Variables
                      Base      Compare
                      DOB         DOB        Diff.       % Diff
    PATNO
    _____        _____   _____   _____    _____

        6  ||     05/13/66   05/13/68   731.0000    31.4544


                  Base Value          Compare Value
    PATNO         GENDER              GENDER
    _____     || __                 __

        4  ||      F                     M
```

Suppose you only want to verify that the genders are correct between the two files. You can add a VAR statement to PROC COMPARE, which restricts the comparison to the variables listed in this statement. Program 7-8 demonstrates this.

Program 7-8   Adding a VAR Statement to PROC COMPARE

```
PROC COMPARE BASE=OLDDEMOG COMPARE=DEMOG BRIEF;
   TITLE "Comparing Demographic Information between Two Data Sets";
   ID PATNO;
   VAR GENDER;
RUN;
```

Here is the output, which now only shows a gender comparison between the two files.

```
Comparing Demographic Information between Two Data Sets

COMPARE Procedure
Comparison of WORK.OLDDEMOG with WORK.DEMOG
(Method=EXACT)

NOTE: Data set WORK.OLDDEMOG contains 2 observations not in
      WORK.DEMOG.
NOTE: Data set WORK.DEMOG contains 1 observations not in
      WORK.OLDDEMOG.
NOTE: Values of the following 1 variables compare unequal: GENDER

                                                       Continued
```

```
Value Comparison Results for Variables

               Base Value              Compare Value
   PATNO       GENDER                  GENDER

   _____    __                      __

       4       F                       M
```

You have seen a few examples of how PROC COMPARE can be used to perform a comparison between two raw data files or two SAS data sets. Before going out and spending a lot of money on custom designed software, you may want to take a look at PROC COMPARE.

8 Some SQL Solutions to Data Cleaning

Introduction

It was a hard decision whether to group all the PROC SQL approaches together in one chapter or to include an SQL solution in each of the other chapters. I opted for the former. PROC SQL (Structured Query Language) is an alternative to the traditional DATA step and PROC approaches used in this book up to this point. PROC SQL is sometimes easier to program and more efficient, sometimes less so, sometimes **extremely** less efficient (as with full joins). In this chapter, many of the data cleaning operations you performed earlier with DATA step and PROC solutions will be revisited. Be careful, some SAS programmers get carried away with PROC SQL and try to solve every problem with it. It is extremely powerful and useful, but it is not always the best solution to every problem. For a more complete discussion of PROC SQL, I recommend the *SAS Guide to the SQL Procedure, Usage and Reference, Version 6*, published by SAS Institute.

A Quick Review of PROC SQL

PROC SQL can be used to list data to the output device (Output window), to create SAS data sets (also called tables in SQL terminology), to create SAS views, or to create macro variables. For most of your data cleaning operations, you will not be creating SAS data sets or views. By omitting the CREATE statement, the results of an SQL query will be sent to the Output window (unless the NOPRINT option is included). For example, if you have a data set called ONE and you want to list all observations where X is greater than 100, you would write a program like the one in Program 8-1.

Program 8-1   Demonstrating a Simple SQL Query

```
PROC SQL;
   SELECT X
   FROM ONE
   WHERE X GT 100;
QUIT;
```

The SELECT statement identifies which variables you want to select. This can be a single variable (as in this example), a list of variables (separated by commas, not spaces), or an asterisk (*), which means all the variables in the data set. The FROM statement identifies what data set to read. Finally, the WHERE statement selects only those observations for which the WHERE statement is true. This is a good time to mention that the SQL statements have to be in a certain order: SELECT, FROM, WHERE, GROUP BY (having), ORDER BY. Thanks to Cynthia Zender (one of my reviewers), you can remember this order by the saying: *S*an *F*rancisco *W*here the *G*rateful Dead *O*riginate.

PROC SQL has one significant advantage over DATA step approaches when combining data sets. With SQL, you can perform a many-to-many merge to produce a Cartesian Product, which contains an observation (row) for every combination of observations in the two data sets. A many-to-many merge in a DATA step does not produce a Cartesian Product, and the result is usually not useful. Many of these operations are demonstrated in the sections that follow.

Checking for Invalid Character Values

Let's start with checking for invalid character values. For these examples, let's use the SAS data set PATIENTS (see the Appendix for the program and data file), and look for invalid values for GENDER, DX, and AE. In Program 8-2, missing values are reported as invalid for GENDER and AE. Later, the program is modified so that missing values are not reported.

Program 8-2   Using SQL to Look for Invalid Character Values

```
LIBNAME CLEAN "C:\CLEANING";
***Checking for invalid character data;
PROC SQL;
   TITLE "Checking for Invalid Character Data";
   SELECT PATNO,
          GENDER,
          DX,
          AE
   FROM CLEAN.PATIENTS
   WHERE GENDER NOT IN ('M','F')          OR
         VERIFY(DX,'0123456789 ') NE 0    OR
         AE NOT IN ('0','1');
QUIT;
```

Because there is no CREATE statement, the observations meeting the WHERE clause will be printed to the Output window when the procedure is submitted. The variables listed in the SELECT statement are separated by commas in PROC SQL, not spaces as in a VAR statement in a DATA step. Notice that the SQL solution looks very much like the DATA step solution used in Chapter 1, which uses the simple way of checking for invalid DX values (see page 9 for details).

Here is the output from running Program 8-2.

```
Checking for Invalid Character Data

Patient              Diagnosis  Adverse
Number     Gender    Code       Event?

002        F         X          0
003        X         3          1
004        F         5          A
006                  6          1
010        f         1          0
013        2         1
002        F         X          0
023        f                    0
```

If you do not want missing values for GENDER and AE to be identified as invalid, include a missing value (blank) in the list of valid values, as shown in Program 8-3.

Program 8-3   Using SQL to List Invalid Character Data: Missing Values Not Flagged as Errors

```
PROC SQL;
   TITLE "Checking for Invalid Character Data";
   TITLE2 "Missing Values Not Flagged as Errors";
   SELECT PATNO,
          GENDER,
          DX,
          AE
   FROM CLEAN.PATIENTS
   WHERE GENDER NOT IN ('M','F',' ')        OR
         VERIFY(DX,'0123456789 ') NE 0      OR
         AE NOT IN ('0','1',' ');
QUIT;
```

Checking for Outliers

A similar program can be used to check for out-of-range numeric values. The SQL statements in Program 8-4 produce a report for heart rate, systolic blood pressure, and diastolic blood pressure readings outside specified ranges. Because missing values are not in the specified ranges, they will be reported as errors by this program.

Program 8-4   Using SQL to Check for Out-of-Range Numeric Values

```
PROC SQL;
   TITLE "Checking for Out-of-Range Numeric Values";
   SELECT PATNO,
          HR,
          SBP,
          DBP
   FROM CLEAN.PATIENTS
   WHERE HR  NOT BETWEEN 40 AND 100        OR
         SBP NOT BETWEEN 80 AND 200        OR
         DBP NOT BETWEEN 60 AND 120;
QUIT;
```

The WHERE statement can be written many ways, just as with a WHERE statement in a DATA step. The output from these statements is shown next.

```
Checking for Out-of-Range Numeric Values

                      Systolic  Diastolic
  Patient    Heart      Blood     Blood
  Number      Rate    Pressure  Pressure

   004         101       200       120
   008         210        .         .
   009          86       240       180
   010          .         40       120
   011          68       300        20
   014          22       130        90
   017         208        .         84
   123          60        .         .
   321         900       400       200
   020          10        20         8
   023          22        34        78
   027          .        166       106
   029          .         .         .
```

If you don't want to consider missing values as errors, make this simple modification to the WHERE statement:

```
WHERE HR  NOT BETWEEN 40 AND 100 AND HR  IS NOT MISSING    OR
      SBP NOT BETWEEN 80 AND 200 AND SBP IS NOT MISSING    OR
      DBP NOT BETWEEN 60 AND 120 AND DBP IS NOT MISSING;
```

The terms IS MISSING or IS NOT MISSING can be used with either character or numeric variables. Also, the term NULL can be used in place of the word MISSING.

Checking a Range Using an Algorithm Based on the Standard Deviation

In Chapter 2, on page 58, an algorithm to detect outliers based on standard deviation was described. Program 8-5 shows an SQL approach using the same algorithm and using the variable systolic blood pressure (SBP).

Program 8-5 **Using SQL to Check for Out-of-Range Values Based on the Standard Deviation**

```
PROC SQL;
   SELECT PATNO,
          SBP
   FROM CLEAN.PATIENTS
   HAVING SBP NOT BETWEEN MEAN(SBP) - 2 * STD(SBP) AND
      MEAN(SBP) + 2 * STD(SBP)                    AND
      SBP IS NOT MISSING;
QUIT;
```

This program uses two summary functions, MEAN and STD. When these functions are used, a HAVING clause is needed instead of the WHERE clause used earlier. In this example, all values more than two standard deviations away from the mean that are not missing are printed to the Output window.

Here is the output after running Program 8-5.

Patient Number	Systolic Blood Pressure
011	300
321	400

The SQL procedure can be made more general by turning the program into a macro and making the variable name a macro variable, as shown in Program 8-6.

Program 8-6   Converting Program 8-5 into a Macro

```
%MACRO RANGESTD(DSN,VARNAME);
   PROC SQL;
      SELECT PATNO,
             &VARNAME
      FROM &DSN
      HAVING &VARNAME NOT BETWEEN MEAN(&VARNAME) - 2 * STD(&VARNAME) AND
         MEAN(&VARNAME) + 2 * STD(&VARNAME)                         AND
         &VARNAME IS NOT MISSING;
   QUIT;
%MEND RANGESTD;
```

For example, here is the statement to call this macro to test the variable DBP.

```
%RANGESTD(CLEAN.PATIENTS,DBP)
```

Here is the corresponding output.

Patient Number	Diastolic Blood Pressure
009	180
321	200
020	8

Checking for Missing Values

It's particularly easy to use PROC SQL to check for missing values. The WHERE clause IS MISSING can be used for both character and numeric variables. The simple query shown in Program 8-7 checks the data set for all character and numeric missing values and prints out any observation that contains a missing value for one or more variables.

Program 8-7   Using SQL to List Missing Values

```
PROC SQL;
   SELECT *
   FROM CLEAN.PATIENTS
   WHERE PATNO   IS MISSING OR
         GENDER  IS MISSING OR
         VISIT   IS MISSING OR
         HR      IS MISSING OR
         SBP     IS MISSING OR
         DBP     IS MISSING OR
         DX      IS MISSING OR
         AE      IS MISSING;
QUIT;
```

The SELECT statement uses an asterisk (*) to indicate that all the variables in the data set are listed in the FROM statement.

Here is the output from Program 8-7.

Patient Number	Gender	Visit Date	Heart Rate	Systolic Blood Pressure	Diastolic Blood Pressure	Diagnosis Code	Adverse Event?
006		06/15/1999	72	102	68	6	1
007	M	.	88	148	102		0
008	F	08/08/1998	210	.	.	7	0
010	f	10/19/1999	.	40	120	1	0
011	M	.	68	300	20	4	1
012	M	10/12/1998	60	122	74		0
013	2	08/23/1999	74	108	64	1	
014	M	02/02/1999	22	130	90		1
003	M	11/12/1999	58	112	74		0
015	F	.	82	148	88	3	1
017	F	04/05/1999	208	.	84	2	0
019	M	06/07/1999	58	118	70		0
123	M	.	60	.	.	1	0
321	F	.	900	400	200	5	1
020	F	.	10	20	8		0
023	f	12/31/1998	22	34	78		0
027	F	.	.	166	106	7	0
029	M	05/15/1998	.	.	.	4	1

Range Checking for Dates

You can also use PROC SQL to check for dates that are out of range. Suppose you want a list of all patients in the PATIENTS data set that have nonmissing visit dates before June 1, 1998 or after October 15, 1999. You use the same programming statements used for checking out-of-range numeric quantities except that you use a date constant in place of the actual number of days from January 1, 1960 (day 0 according to SAS). The SQL statements in Program 8-8 will do the trick.

Program 8-8   Using SQL to Perform Range Checks on Dates

```
PROC SQL;
   TITLE "Dates Before June 1, 1998 or After October 15, 1999";
   SELECT PATNO,
          VISIT
   FROM CLEAN.PATIENTS
   WHERE VISIT NOT BETWEEN '01JUN1998'D AND '15OCT1999'D AND
         VISIT IS NOT MISSING;
QUIT;
```

Here is the resulting output from Program 8-8.

```
Dates Before June 1, 1998 or After October 15, 1999

Patient
Number   Visit Date

XX5      05/07/1998
010      10/19/1999
003      11/12/1999
028      03/28/1998
029      05/15/1998
```

Checking for Duplicates

In Chapter 5, you used PROC SORT with the NODUP and NODUPKEY options to detect duplicates, as well as a DATA step approach using the FIRST. and LAST. temporary variables. Yes, you guessed it, there is an SQL answer also. If you have a GROUP BY statement in your PROC SQL and follow it with a COUNT function, you can count the frequency of each level of the GROUP BY variable. (COUNT is the SQL name for the N and FREQ functions in the SAS System that count the number of nonmissing arguments.) If you choose patient number (PATNO) as the grouping variable, the COUNT function will tell you how many observations there are per patient. Remember to use a HAVING clause when you use summary functions such as COUNT. Look at the SQL code in Program 8-9.

Program 8-9   Using SQL to List Duplicate Patient Numbers

```
PROC SQL;
   TITLE "Duplicate Patient Numbers";
   SELECT PATNO,
          VISIT
     FROM CLEAN.PATIENTS
     GROUP BY PATNO
     HAVING COUNT(PATNO) GT 1;
QUIT;
```

In Program 8-9, you are telling PROC SQL to list any duplicate patient numbers. Note that multiple missing patient numbers will not appear in the listing because the COUNT function returns a frequency count only for nonmissing values. Here are the results of running Program 8-9.

```
Duplicate Patient Numbers

Patient
Number    Visit Date
─────────────────────
002       11/13/1998
002       11/13/1998
003       10/21/1998
003       11/12/1999
006       06/15/1999
006       07/07/1999
```

Identifying Subjects with "n" Observations Each

Using the grouping capability of PROC SQL and the COUNT function, you can list all patients that do not have exactly "n" visits or observations in a data set, just as you did in Programs 5-11 and 5-12. Here is the program with an explanation following.

Program 8-10 Using SQL to List Patients Who Do Not Have Two Visits

```
TITLE "Listing of Patients Who Do Not Have Two Visits";
PROC SQL;
   SELECT PATNO,
          VISIT
      FROM CLEAN.PATIENTS2
      GROUP BY PATNO
      HAVING COUNT(PATNO) NE 2;
QUIT;
```

By first grouping the observations by patient number, you can then use the COUNT function, which returns the number of observations in a group. Here is the output from Program 8-10.

```
Listing of Patients Who Do Not Have Two Visits

PATNO      VISIT

002     01/01/1999
002     01/10/1999
002     02/09/1999
003     10/21/1998
006     11/11/1998
```

Checking for an ID in Each of Two Files

Do you think PROC SQL can check if each patient number is in two files? Why else is there a section heading with that task listed? Of course you can! Now, on to the problem.

The equivalent of a DATA step merge is called a JOIN in SQL terms. Normally, a JOIN lists only those observations that have a matching value for the variables in each of the files. If you want all observations from both files, regardless if they have a corresponding observation in the other file, you perform a FULL JOIN (this is equivalent to a MERGE where no IN= variables are used). So, if you perform a FULL JOIN between two data sets and an ID value is not in both data sets, one of the observations will have a missing value for the ID variable. Let's use the same data sets, ONE and TWO, that were used in Chapter 6. For convenience, the code to produce these data sets is shown in Program 8-11.

Program 8-11 Creating Two Data Sets for Testing Purposes

```
DATA ONE;
   INPUT PATNO X Y;
DATALINES;
1 69 79
2 56 .
3 66 99
5 98 87
12 13 14
;
DATA TWO;
   INPUT PATNO Z;
DATALINES;
1 56
3 67
4 88
5 98
13 99
;
```

Program 8-12 shows the SQL program.

Program 8-12 Using SQL to Look for ID's That Are Not in Each of Two Files

```
PROC SQL;
   TITLE "Patient Numbers Not in Both Files";
   SELECT ONE.PATNO AS ID_ONE,
          TWO.PATNO AS ID_TWO
   FROM ONE FULL JOIN TWO
   ON ONE.PATNO EQ TWO.PATNO
   WHERE ONE.PATNO IS MISSING OR TWO.PATNO IS MISSING;
QUIT;
```

Because the variable name PATNO is used in both data sets, you can distinguish between them by adding either ONE. or TWO. before the variable name, depending on whether you are referring to the patient number from data set ONE or data set TWO. Also, to make it easier to keep track of these two variables, an alias (created by using the AS statement) for each of these variable names was created (ID_ONE and ID_TWO).

The condition for the FULL JOIN is that the ID's match between the two data sets. This is specified in the ON statement. (Note: You can have a different variable name for the ID variable in each file without causing any problems. In the DATA step solution, using a MERGE statement, you need to rename one of the ID variables so that they are the same.) The WHERE clause will be true for any ID that is not in both data sets.

The output from this query contains all the observations where an ID value is not in both files, as shown next.

```
Patient Numbers Not in Both Files

   ID_ONE                ID_TWO
  _____              _____

     2                     .
     .                     4
    12                     .
     .                    13
```

More Complicated Multi-File Rules

Let's start the discussion of more complicated multi-file rules by redoing the example from Chapter 6. To review, there are two files: AE, which recorded adverse events for patients in the study, and LAB_TEST, which contained the laboratory tests for people with various adverse events (see the Appendix for the programs to create these data sets). The goal is to list any patient who had an adverse event of 'X' anywhere in the adverse event data set who either did not have any entry in the laboratory data set or where the date of the lab test was before the date of the adverse event. The following SQL query will produce the same information as the DATA step solution shown in Chapter 6 in Program 6-8.

Program 8-13 Using SQL to Demonstrate More Complicated Multi-File Rules

```
PROC SQL;
    TITLE1 "Patients with an AE of X Who Did Not Have a";
    TITLE2 "Labtest or Where the Date of the Test Is Prior";
    TITLE3 "to the Date of the Visit";
    SELECT AE.PATNO AS AE_PATNO LABEL="AE Patient Number",
           A_EVENT,
           DATE_AE,
           LAB_TEST.PATNO AS LABPATNO LABEL="LAB Patient Number",
           LAB_DATE
    FROM CLEAN.AE LEFT JOIN CLEAN.LAB_TEST
    ON AE.PATNO=LAB_TEST.PATNO
    WHERE A_EVENT = 'X'            AND
          LAB_DATE LT DATE_AE;
QUIT;
```

Because the variable PATNO had the same label in both the AE and LAB_TEST data sets, a LABEL column modifier was used to re-label these variables so that they could be distinguished in the output listing. An alias (AE_PATNO and LABPATNO), as well as a label, was selected for each of these variables.

To help explain the difference between a LEFT JOIN, a RIGHT JOIN, and a FULL JOIN, let's execute all three with the data sets ONE and TWO, which were described in the previous section. In Program 8-14, the following SQL statements execute all three joins.

Program 8-14 Example of LEFT, RIGHT, and FULL Joins

```
PROC SQL;
    TITLE "Left Join";
    SELECT ONE.PATNO AS ONE_ID,
           TWO.PATNO AS TWO_ID
    FROM ONE LEFT JOIN TWO
    ON ONE.PATNO EQ TWO.PATNO;
```

```
    TITLE "Right Join";
    SELECT ONE.PATNO AS ONE_ID,
           TWO.PATNO AS TWO_ID
    FROM ONE RIGHT JOIN TWO
    ON ONE.PATNO EQ TWO.PATNO;

    TITLE "Full Join";
    SELECT ONE.PATNO AS ONE_ID,
           TWO.PATNO AS TWO_ID
    FROM ONE FULL JOIN TWO
    ON ONE.PATNO EQ TWO.PATNO;
QUIT;
```

By inspecting the next three listings, it is very easy to see the difference among these three different JOIN operations.

```
Left Join

    ONE_ID     TWO_ID

        1          1
        2          .
        3          3
        5          5
       12          .

Right Join

    ONE_ID     TWO_ID

        1          1
        3          3
        .          4
        5          5
        .         13

Full Join

    ONE_ID     TWO_ID

        1          1
        2          .
        3          3
        .          4
        5          5
       12          .
        .         13
```

Now, back to our example. You want all patients in the AE data set with an adverse event equal to 'X', but only those observations in the LAB_TEST data set where the patient numbers were selected from the adverse event data set. Therefore, the LEFT JOIN operation was used.

Finally, if the lab date (LAB_DATE) is prior to the adverse event or if the lab date is missing, the statement LAB_DATE LT DATE_AE will be true. Notice, in the output that follows, that the same three patients are listed with this query as were listed in the example in Chapter 6.

```
Patients with an AE of X Who Did Not Have a
Labtest or Where the Date of the Test Is Prior
to the Date of the Visit

AE                              LAB
Patient  Adverse                Patient     Date of
Number   Event    Date of AE    Number      Lab Test
-------------------------------------------------------
009      X        12/25/1998                       .
011      X        10/10/1998                       .
025      X        02/09/1999    025      01/01/1999
```

PROC SQL provides a very convenient way to conduct many of the data cleaning tasks described throughout this book. There are those of us who still feel more comfortable with DATA step and PROC approaches and others who feel that PROC SQL is the solution to all their problems. You may want to start using PROC SQL more often and become more comfortable with its syntax. Quite likely, you will wind up using a combination of traditional DATA step and PROC approaches along with PROC SQL for your data cleaning and other SAS programming needs.

9 Using Validation Data Sets

Introduction

In previous chapters, SAS programs and macros were used to check for invalid data values. The rules were "hard-coded" or entered as calling arguments to SAS macros. In this chapter, you will see how data rules can be entered into a raw data file and turned into a SAS validation data set, with the data checking operation performed by a general purpose cleaning/validation program. This approach allows you to create permanent validation data sets, containing rules for frequently used variables, and to apply these validation data sets against any data set that needs to be validated. Finally, an introduction to SAS Integrity Constraints is presented. Integrity constraints allow you to store rules concerning your variables in the data set itself. This feature is available starting with Version 7 of SAS software.

A Simple Example of a Validation Data Set

Let's start with a simple validation data set that only handles range checking for numeric variables. The following three rules are used for this example.

Valid heart rate values are between 40 and 100.
Valid values for systolic blood pressure are between 80 and 200.
Valid values for diastolic blood pressure are between 60 and 120.

For this simple example, all missing values are treated as invalid. Program 9-3 is generalized to treat missing values as either valid or invalid for each numeric variable.

For this example, the validation data set contains the variable name, the minimum valid value, and the maximum valid value.

Program 9-1 creates a validation data set called VALID.

Program 9-1   Creating a Simple Validation Data Set

```
DATA VALID;
    INFILE "C:\CLEANING\VALID1.TXT" MISSOVER;
    INPUT VARNAME : $32. MIN MAX;    ❶
    VARNAME = UPCASE(VARNAME);
RUN;
```

where the data file VALID1.TXT contains the following:

```
HR 40 100
SBP 80 200
DBP 60 140
```

Notice that the program is allowing variable names up to 32 characters in length ❶ (Version 7 and later), and that the UPCASE function is used to make sure all the variable names are in uppercase.

The first step in applying this validation data set against the PATIENTS data set is to restructure the patient data. For each patient ID, you want a separate observation for each numeric variable in the data set. You also want to create two new variables, VARNAME and VALUE, which will hold the name of the numeric variable and its value, respectively. To understand what this structure looks like, here is what the first 10 observations in this restructured data should look like.

```
First 10 Observations in the Restructured Patients Data Set

    Patient ID          Variable Name              Value
     (PATNO)              (VARNAME)                (VALUE)

        001                  VISIT                  14194
        001                  HR                        88
        001                  SBP                      140
        001                  DBP                       80
        002                  VISIT                  14196
        002                  HR                        84
        002                  SBP                      120
        002                  DBP                       78
        003                  VISIT                  14173
        003                  HR                        68
```

As you may have guessed at this point, you want this particular structure so that it can be merged with the VALID data set, the goal being to add the minimum and maximum cutoffs to each observation in this restructured data set. Program 9-2 will perform this restructuring task.

Note: Even though the task at hand is to validate the three numeric variables HR, SBP, and DBP, Program 9-2 was written to be more general and to include all the numeric variables in the data set.

Program 9-2   Restructuring the PATIENTS Data Set and Producing an Exceptions Report

```
***Restructure PATIENTS;
DATA PAT;
   SET CLEAN.PATIENTS;
   ***Make room for variable names up to 32 characters;
   LENGTH VARNAME $ 32;
   ***Array to contain all numeric variables;
   ARRAY NUMS[*] _NUMERIC_;   ❶

   DO I = 1 TO DIM(NUMS);   ❷
      CALL VNAME(NUMS[I],VARNAME);   ❸
      VARNAME = UPCASE(VARNAME);   ❹
      VALUE = NUMS[I];   ❺
      OUTPUT;   ❻
   END;
   KEEP PATNO VARNAME VALUE;
RUN;
```

The keyword _NUMERIC_, used in the ARRAY statement ❶, creates the NUMS array with all the numeric variables in the data set CLEAN.PATIENTS as its elements. The DIM function ❷ returns the number of elements in the array. The key to this program is the CALL VNAME statement ❸. This very useful function takes as its first argument an element of an array, and places the associated variable name in the second argument (VARNAME in this case) ❹. It is important to set the length of the character variable that is to hold the variable name, earlier in the DATA step. A length of 32 was chosen to be compatible with Versions 7 and later of SAS software. For the same reason that the UPCASE function was used with the VALID data set, it is also used here to ensure that both data sets contain the variable names in the same case (so they can be merged). Finally, VALUE ❺ is assigned the numeric value of the variable. Because the OUTPUT statement ❻ is within the DO loop, there will be as many observations per patient as there are numeric variables in the original data set. The next step is to sort both this data set (PAT) and the VALID data set by VARNAME.

Program 9-2   Restructuring   the   PATIENTS   Data   Set   and   Producing   an Exceptions Report (Continued)

```
PROC SORT DATA=PAT;
   BY VARNAME;
RUN;

PROC SORT DATA=VALID;
   BY VARNAME;
RUN;
```

Let's merge the two files.

Program 9-2   Restructuring   the   PATIENTS   Data   Set   and   Producing   an Exceptions Report (Continued)

```
DATA VERIFY;
   MERGE PAT(IN=IN_PAT)
         VALID(IN=IN_VALID);
   BY VARNAME;
   IF IN_PAT AND IN_VALID AND
      (VALUE LT MIN OR VALUE GT MAX);    ❼
RUN;
```

Because you only want information on those variables that are in both data sets, each of the data sets in the MERGE statement is followed by the IN= data set option. The subsetting IF statement ❼ accomplishes this. In addition, this subsetting IF statement is used to check if the value of your variable is outside the acceptable range. In this example, missing values, which are logically less than the minimum, are included in the list of invalid values. The program is generalized in the next section. The last step is to sort the merged data set by patient number and variable name, so that an exception report, in patient number order, can be produced.

Program 9-2   Restructuring the PATIENTS Data Set and Producing an Exceptions Report (Continued)

```
PROC SORT DATA=VERIFY;
   BY PATNO VARNAME;
RUN;

PROC PRINT DATA=VERIFY;
   TITLE "Exceptions Report";
   ID PATNO;
   VAR VARNAME VALUE;
RUN;
```

The output from the above PROC PRINT is shown next.

```
Exceptions Report

PATNO     VARNAME      VALUE

 004        HR          101
 008        DBP          .
 008        HR          210
 008        SBP          .
 009        DBP         180
 009        SBP         240
 010        HR           .
 010        SBP          40
 011        DBP          20
 011        SBP         300
 014        HR           22
 017        HR          208
 017        SBP          .
 020        DBP           8
 020        HR           10
 020        SBP          20
 023        HR           22
 023        SBP          34
 027        HR           .
 029        DBP          .
 029        HR           .
 029        SBP          .
 123        DBP          .
 123        SBP          .
 321        DBP         200
 321        HR          900
 321        SBP         400
```

If you do not want to list missing values as exceptions, you can substitute the following subsetting IF statement in Program 9-2:

```
IF IN_PAT AND IN_VALID AND
      ((VALUE LT MIN AND VALUE NE . ) OR VALUE GT MAX);
```

Making the Program More Flexible and Converting It to a Macro

Now that you have the general idea under your belt, let's make the program more flexible by treating missing values as valid or invalid for each of the numeric variables. At the same time, let's also turn the program into a macro.

In the file containing the acceptable ranges for each of the variables, you are going to add a "flag" to indicate if missing values are okay or if they should be treated as data errors. The default will be to treat missing values as errors. To override this default, a new variable (MISS_OK) will be added to the VALID data set. Values of 'Y' indicate that missing values are okay, anything else (including nothing) indicates that you want the default behavior of the program. To demonstrate this new macro, let's create a validation data set with the following new set of rules:

- Valid heart rate values are between 40 and 100. Missing values are not valid.
- Valid values for systolic blood pressure are between 80 and 200. Missing values are valid.
- Valid values for diastolic blood pressure are between 60 and 120. Missing values are not valid.

Following these rules, the new program (Program 9-3) creates the VALID data set.

Program 9-3   Creating a New Validation Data Set That Contains Missing Value Instructions

```
PROC FORMAT;
    INVALUE $MISS(UPCASE DEFAULT=1) 'Y' = 'Y'
                                    OTHER = 'N';
RUN;


***Create a validation data set from the raw data;
DATA VALID;
    INFILE "C:\CLEANING\VALID2.TXT" MISSOVER;   ❶
    LENGTH VARNAME $ 32 MISS_OK $ 1;
    INPUT VARNAME $ MIN MAX MISS_OK : $MISS.;
    ***Make sure all variable names are in uppercase so they
       will match the variable names in the data set to be checked;
    VARNAME = UPCASE(VARNAME);
RUN;
```

Where the data file VALID2.TXT looks like this:

```
hr 40 100
sbp 80 200 y
dbp 60 140 n
```

Notice the MISSOVER option in the INFILE statement ❶. This is critical with list-directed (spaces between the values) data. With this option, if the INPUT statement runs out of data before it runs out of variables, it assigns a missing value for any of the remaining variables. Also, notice the user-defined informat $MISS. This informat converts all character data to uppercase (by the UPCASE option), and it converts any character other than an uppercase or lowercase 'Y' to the value of 'N'. Thus, both heart rate (HR) and diastolic blood pressure (DBP) will follow the default behavior of treating missing values as invalid while missing values for systolic blood pressure (SBP) will not be listed as errors.

The macro Program 9-4, is similar to Program 9-2. It is called with three arguments; the name of the ID variable, the name of the data set to be validated, and the name of the validation data file. The code to create the validation data set from the raw validation data file is also included as part of the macro.

Note: Some variable names in this program and other programs in this chapter are longer than eight characters in length. They need to be shortened if you are using pre-Version 7 SAS software.

Program 9-4   Validating a Data Set with a Macro That Contains Missing Value Instructions

```
*-------------------------------------------------------------------*
| Program Name: VALID_NUM.SAS  in C:\CLEANING                       |
| Purpose: Macro that takes an ID variable, a SAS data set to be    |
|          validated, and a validation data file, and prints an     |
|          exception report to the output device.                   |
|          This macro is for numeric variable range checking only.  |
| Arguments: ID     - ID variable                                   |
|            DATASET - SAS data set to be validated                 |
|            VALID_FILE - Validation data file                      |
|                     Each line of this file contains the name      |
|                     of a numeric variable, the minimum value,     |
|                     the maximum value, and a missing value        |
|                     indicator ('Y' missing values OK, 'N'         |
|                     missing values not OK), all separated by      |
|                     at least one space.                           |
|Example:  %VALID_NUM(PATNO,CLEAN.PATIENTS,C:\CLEANING\VALID2.TXT)  |
*-------------------------------------------------------------------*;
```

```
%MACRO VALID_NUM (ID,                /* ID variable                     */
                  DATASET,           /* Data set to be validated        */
                  VALID_FILE,        /* Validation data set             */
                  );

   ***Note: For pre-Version 7, substitute a shorter name
      for several variables;

   ***Create the validation data set;
   PROC FORMAT;
      INVALUE $MISS(UPCASE DEFAULT=1) 'Y' = 'Y'
                                      OTHER = 'N';
   RUN;

   DATA VALID;
      INFILE "&VALID_FILE" MISSOVER;
      LENGTH VARNAME $ 32 MISS_OK $ 1;
      INPUT VARNAME $ MIN MAX MISS_OK : $MISS.;
      VARNAME = UPCASE(VARNAME);
   RUN;

   ***Restructure &DATASET;
   DATA PAT;
      SET &DATASET;
      LENGTH VARNAME $ 32;
      ARRAY NUMS[*] _NUMERIC_;
      N_NUMS = DIM(NUMS);
      DO I = 1 TO N_NUMS;
         CALL VNAME(NUMS[I],VARNAME);
         VARNAME = UPCASE(VARNAME);
         VALUE = NUMS[I];
         OUTPUT;
      END;
      KEEP PATNO VARNAME VALUE;
   RUN;

   PROC SORT DATA=PAT;
      BY VARNAME PATNO;
   RUN;

   PROC SORT DATA=VALID;
      BY VARNAME;
   RUN;
```

```
***Merge the validation data set and the restructured SAS data set;
DATA VERIFY;
   MERGE PAT(IN=IN_PAT) VALID(IN=IN_VALID);
   BY VARNAME;
   IF (IN_PAT AND IN_VALID)   AND
      (VALUE LT MIN OR VALUE GT MAX)   AND
      NOT(VALUE = . AND MISS_OK EQ 'Y');
RUN;

PROC SORT DATA=VERIFY;
   BY PATNO VARNAME;
RUN;

***Reporting section;
OPTIONS NODATE NONUMBER;
TITLE;

DATA _NULL_;
   FILE PRINT HEADER = REPORT_HEAD;
   SET VERIFY;
   BY PATNO;

   IF VALUE = . THEN PUT
      @1   PATNO
      @18 VARNAME
      @39 "Missing";

   ELSE IF VALUE GT . AND VALUE LT MIN THEN PUT
      @1   PATNO
      @18 VARNAME
      @29 VALUE
      @39 "Below Minimum ("  MIN  +(-1) ")";

   ELSE IF VALUE GT MAX THEN PUT
      @1   PATNO
      @18 VARNAME
      @29 VALUE
      @39 "Above Maximum ("  MAX  +(-1) ")";

  IF LAST.PATNO THEN PUT ;
     RETURN;
```

```
REPORT_HEAD:
   PUT @1 "Exceptions Report for Data Set &DATASET" /
          "Using Validation Data File &VALID_FILE"  //
        @1  "Patient ID"
        @18 "Variable"
        @29 "Value"
        @39 "Reason" /
        @1  60*"-";
RUN;

***Cleanup temporary data sets;
PROC DATASETS LIBRARY=WORK NOLIST;
   DELETE PAT;
   DELETE VERIFY;
RUN;
QUIT;

%MEND VALID_NUM;
```

Notice that the simple PROC PRINT was replaced with a nicer looking exception report by using PUT statements. Executing the macro on the PATIENTS data set by using this statement

```
%VALID_NUM(PATNO,CLEAN.PATIENTS,C:\CLEANING\VALID2.TXT)
```

produces the output shown next.

```
Exceptions Report for Data Set CLEAN.PATIENTS
Using Validation Data File C:\CLEANING\VALID2.TXT

Patient ID      Variable   Value    Reason
-------------------------------------------------------------
004             HR         101      Above Maximum (100)

008             DBP                 Missing
008             HR         210      Above Maximum (100)

009             DBP        180      Above Maximum (140)
009             SBP        240      Above Maximum (200)

010             HR                  Missing
010             SBP        40       Below Minimum (80)

011             DBP        20       Below Minimum (60)
011             SBP        300      Above Maximum (200)
014             HR         22       Below Minimum (40)

017             HR         208      Above Maximum (100)

020             DBP        8        Below Minimum (60)
020             HR         10       Below Minimum (40)
020             SBP        20       Below Minimum (80)

023             HR         22       Below Minimum (40)
023             SBP        34       Below Minimum (80)

027             HR                  Missing

029             DBP                 Missing
029             HR                  Missing

123             DBP                 Missing

321             DBP        200      Above Maximum (140)
321             HR         900      Above Maximum (100)
321             SBP        400      Above Maximum (200)
```

Notice that the missing values for systolic blood pressure (SBP) are not listed as errors because of the value of 'Y' for the missing value indicator (see data file VALID2.TXT).

Validating Character Data

You can use similar techniques to validate character data. Here, the set of rules might contain a list of valid values for each character variable and instructions on how to treat missing values. To do your testing, create a new data set (TEST_CHAR) by running Program 9-5.

Program 9-5   Creating a Test Data Set for Character Validation

```
DATA TEST_CHAR;
   ***If there is a short record, set all variables to missing
      using the MISSOVER option in the INFILE statement.
      DATALINES is a special file reference that allows INFILE
      statement options to be used with data following a
      DATALINES statement, rather than an external file;
   INFILE DATALINES MISSOVER;
   LENGTH PATNO $ 3 CODE $ 2 GENDER  AE $ 1;
   INPUT PATNO CODE GENDER AE;
DATALINES;
001 A M 0
002 AB F 1
003 BA F .
004 . . .
005 X Y Z
006 AC M 0
;
```

Data set TEST_CHAR contains the variables PATNO, CODE, GENDER, and AE (adverse event). The listing of this data set is shown next.

```
Listing of TEST_CHAR

PATNO     CODE     GENDER     AE

 001       A         M         0
 002       AB        F         1
 003       BA        F
 004
 005       X         Y         Z
 006       AC        M         0
```

Let's include a check for missing values in the program. For this example, assume that valid values for CODE are 'A', 'AB', 'C', and 'D'; valid codes for GENDER are 'F' and 'M'; valid codes for AE are '0' and '1'. Missing values are not valid for GENDER but are valid for CODE and AE.

A three-line data file (called a validation data file) will be used to store these rules. Each line of this file will hold the rules for one variable. Each line begins with a variable name. Following one or more spaces, you enter one or more valid values for this variable. End this list with at least two spaces followed by a 'Y' (missing values are okay) or an 'N' (missing values are not okay). Actually, you can leave off the 'N' if you want to. Any value other than a 'Y', including a missing value, will mean that missing values are not okay. A file written to conform to the rules in the preceding paragraph is shown next.

```
Data file "C:\CLEANING\VALID_C.TXT"

CODE A B AB C    Y
GENDER M F   N
AE 0 1    Y
```

Notice there are at least two spaces between the list of valid character values and the missing value indicator. Program 9-6 creates a validation data set from this validation data file.

Program 9-6   Creating a Validation Data Set (C_VALID) for Character Variables

```
PROC FORMAT;
   INVALUE $MISS (UPCASE DEFAULT=1) 'Y' = 'Y'
                                   OTHER = 'N';
RUN;

DATA C_VALID;
   LENGTH VARNAME $ 32 VALUES_LIST $ 200 MISS_OK $ 1;
   INFILE "C:\CLEANING\VALID_C.TXT" MISSOVER;
   INPUT VARNAME VALUES_LIST & $200. MISS_OK : $MISS.;   ❶
RUN;
```

Let's use the same user-defined informat in this program as was used in Program 9-4. The $MISS. informat turns lowercase and uppercase Y's into uppercase Y's and all other nonmissing values into uppercase N's. The variable called VALUES_LIST is a list of valid values for each of the variables. Notice the use of the ampersand (&) informat modifier in the INPUT statement ❶.

The & modifier changes the default single space delimiter in a list-directed INPUT statement to two or more spaces. It is very important, therefore, to have at least two spaces between the list of valid character values and the missing value indicator (MISS_OK).

The next steps are similar to Program 9-4. First, restructure the PATIENTS data set as before. This time, the reserved name _CHARACTER_ is used instead of _NUMERIC_, so that the CHARS array will contain all the character variables in the data set PATIENTS. Next, sort both data sets prior to the merge step as shown in Program 9-7.

Program 9-7   Writing the Program to Validate Character Variables

```
***Restructure TEST_CHAR;
DATA PAT;
   SET TEST_CHAR;
   ARRAY CHARS[*] _CHARACTER_;
   LENGTH VARNAME $ 32;
   N_CHARS = DIM(CHARS);
   DO I = 1 TO N_CHARS;
      CALL VNAME(CHARS[I],VARNAME);
      VARNAME = UPCASE(VARNAME);
      VALUE = CHARS[I];
      OUTPUT;
   END;
   KEEP PATNO VARNAME VALUE;
RUN;

PROC SORT DATA=PAT;
   BY VARNAME;
RUN;

PROC SORT DATA=C_VALID;
   BY VARNAME;
RUN;
```

In this next section of the program, you need to extract each of the valid character values from the string VALUES_LIST.

Program 9-7   Writing the Program to Validate Character Variables (continued)

```
DATA VERIFY;
   MERGE PAT(IN=IN_PAT) C_VALID(IN=IN_C_VALID);
   BY VARNAME;
   IF (IN_PAT AND IN_C_VALID);   ❶

   LENGTH TOKEN $ 8;

   ***Obviously bad values;
   IF VERIFY (VALUE,VALUES_LIST) NE 0      OR
      VALUE = ' ' AND MISS_OK NE 'Y' THEN DO;   ❷
      OUTPUT;
      RETURN;
   END;

   FLAG = 0;
   DO I = 1 TO 99;
      TOKEN = SCAN(VALUES_LIST,I," ");   ❸
      IF VALUE = TOKEN THEN FLAG + 1;
      IF TOKEN = ' ' OR FLAG > 0 THEN LEAVE;
   END;

   IF FLAG = 0 THEN OUTPUT;
   DROP I TOKEN;
RUN;

PROC SORT DATA=VERIFY;
   BY PATNO VARNAME;
RUN;

***Reporting section;
OPTIONS NODATE NONUMBER;
TITLE;

DATA _NULL_;
   FILE PRINT HEADER = REPORT_HEAD;
   SET VERIFY;
   BY PATNO;
```

```
       IF VALUE = ' '  THEN PUT
          @1   PATNO
          @18  VARNAME
          @39  "Missing";

        ELSE   PUT
          @1   PATNO
          @18  VARNAME
          @29  VALUE
          @39  "Not Valid";

        IF LAST.PATNO THEN PUT ;
        RETURN;

REPORT_HEAD:
     PUT @1  "Exceptions Report for Data Set TEST_CHAR" /
            "Using Validation Data Set VALID_C"  //
          @1   "Patient ID"
          @18  "Variable"
          @29  "Value"
          @39  "Reason" /
          @1   60*"-";
RUN;
```

First, merge the two data sets by VARNAME (the name of the variable to test). Also, subset the data set with the same IF statement used before ❶. For the sake of efficiency, it's best to do a quick test of the entire list of valid values. If the variable to be tested contains any characters that are not anywhere in the valid list, the VERIFY function will return a nonzero value ❷. In the same IF statement, you can check if you have a missing value and the MISS_OK indicator is not a 'Y' (an error). Note that a character value can pass the VERIFY test yet still be invalid. For example, the value 'BA' for patient number 003 returns a 0 from the VERIFY function (it does not contain any invalid characters), but the value of 'BA' is not in the list of valid codes. In the next section of the program, each character value is compared against each of the valid values in the VALUES_LIST.

The SCAN function extracts "words" from a character string. The first argument to the SCAN function ❸ is the character string to be parsed (taken apart) and the second argument is a number which indicates which "word" you want. The last argument is your choice of a delimiter, a blank in this example. Before you enter the DO loop, set a flag equal to 0. If the value of the character variable being tested matches a value in the values list, increment the flag.

As soon as you find a match in the VALUES_LIST or the SCAN function returns a null string (there are no more "words" in the VALUES_LIST), leave the loop. If you finish looping through the valid character values without incrementing the flag, you know that you have a character value that is not in the valid list and you need to output an observation to the VERIFY data set.

It's time now to sort (by patient number and variable name) and produce a report. This code is mostly the same as the previous numeric example given in Program 9-4.

```
PROC SORT DATA=VERIFY;
   BY PATNO VARNAME;
RUN;

***Reporting section;
OPTIONS NODATE NONUMBER;
TITLE;

DATA _NULL_;
   FILE PRINT HEADER = REPORT_HEAD;
   SET VERIFY;
   BY PATNO;

   IF VALUE = ' ' THEN PUT
      @1   PATNO
      @18 VARNAME
      @39 "Missing";

   ELSE  PUT
      @1   PATNO
      @18 VARNAME
      @29 VALUE
      @39 "Not Valid";

   IF LAST.PATNO THEN PUT ;
   RETURN;

REPORT_HEAD:

  PUT @1 "Exceptions Report for Data Set &DATASET" /
         "Using Validation Data Set &VALID"  //
      @1   "Patient ID"
      @18 "Variable"
      @29 "Value"
      @39 "Reason" /
      @1   60*"-";
RUN;
```

The exception report created by this program looks like this.

```
Exceptions Report for Data Set TEST_CHAR
Using Validation Data Set VALID_C

Patient ID       Variable   Value    Reason
-----------------------------------------------------------
003              CODE       BA       Not Valid

004              GENDER              Missing

005              AE         Z        Not Valid
005              CODE       X        Not Valid
005              GENDER     Y        Not Valid

006              CODE       AC       Not Valid
```

As before, let's convert this program into a general purpose macro. This macro includes the program to create the validation data set from a raw data file.

Converting Program 9-7 into a General Purpose Macro

The next step is to convert the program to validate character values into a macro. This is a straightforward task. The same calling arguments as the numeric macro in Program 9-4 are used in Program 9-8.

Program 9-8   Writing a Macro to Check for Invalid Character Values

```
*-------------------------------------------------------------------*
| Program Name: VALID_CHAR in C:\CLEANING                           |
| Purpose: This macro takes an ID variable, a SAS data set to       |
|          be validated, and a validation data file and checks      |
|          for invalid character values and prints an exception     |
|          report.                                                  |
| Arguments: ID        - ID variable name                           |
|            DATASET    - SAS data set to be validated              |
|            VALID_FILE - Validation data file                      |
|                         Each line of this file contains the name  |
|                         of a character variable, a list of valid  |
|                         values separated by spaces, and a 'Y'     |
|                         (missing values okay) or an 'N' (missing  |
|                         values not okay) separated from the list  |
|                         of valid values by 2 or more spaces.      |
| Example: %VALID_CHAR(PATNO,TEST_CHAR,C:\CLEANING\VALID_C.TXT)     |
*-------------------------------------------------------------------*;
%MACRO VALID_CHAR (ID,          /* ID variable               */
                   DATASET,     /* Data set to be validated  */
                   VALID_FILE,  /* Validation data file      */
                   );

   ***Note: For pre-Version 7, substitute a shorter name
      for VALID_FILE;

   ***Create validation data set;
   PROC FORMAT;
      INVALUE $MISS(UPCASE DEFAULT=1) 'Y' = 'Y'
                                      OTHER = 'N';
   RUN;

   PROC FORMAT;
      INVALUE $MISS(UPCASE DEFAULT=1) 'Y' = 'Y'
                                      OTHER = 'N';
   RUN;

   DATA VALID;
      LENGTH VARNAME $ 32 VALUES_LIST $ 200 MISS_OK $ 1;
      INFILE "&VALID_FILE" MISSOVER;
      INPUT VARNAME VALUES_LIST & $200. MISS_OK : $MISS.;
   RUN;
```

```
***Restructure &DATASET;
DATA PAT;
   SET &DATASET;
   ARRAY CHARS[*] _CHARACTER_;
   LENGTH VARNAME $ 32;
   N_CHARS = DIM(CHARS);
   DO I = 1 TO N_CHARS;
      CALL VNAME(CHARS[I],VARNAME);
      VARNAME = UPCASE(VARNAME);
      VALUE = CHARS[I];
      OUTPUT;
   END;
   KEEP &ID VARNAME VALUE;
RUN;

PROC SORT DATA=PAT;
   BY VARNAME;
RUN;

PROC SORT DATA=VALID;
   BY VARNAME;
RUN;

DATA VERIFY;
   MERGE PAT(IN=IN_PAT) VALID(IN=IN_VALID);
   BY VARNAME;
   IF (IN_PAT AND IN_VALID);

   LENGTH TOKEN $ 8;

   ***Obviously bad values;
   IF VERIFY (VALUE,VALUES_LIST) NE 0     OR
      VALUE = ' ' AND MISS_OK NE 'Y' THEN DO;
      OUTPUT;
      RETURN;
   END;

   FLAG = 0;
   DO I = 1 TO 99;
      TOKEN = SCAN(VALUES_LIST,I," ");
      IF VALUE = TOKEN THEN FLAG + 1;
      IF TOKEN = ' '  OR FLAG > 0 THEN LEAVE;
   END;
```

```
   IF FLAG = 0 THEN OUTPUT;
      DROP I TOKEN;
   RUN;

   PROC SORT DATA=VERIFY;
      BY &ID VARNAME;
   RUN;

   ***Reporting section;
   OPTIONS NODATE NONUMBER;
   TITLE;

   DATA _NULL_;
      FILE PRINT HEADER = REPORT_HEAD;
      SET VERIFY;
      BY &ID;

      IF VALUE = ' ' THEN PUT
         @1  &ID
         @18 VARNAME
         @39 "Missing";

      ELSE  PUT
         @1  &ID
         @18 VARNAME
         @29 VALUE
         @39 "Not Valid";

      IF LAST.&ID THEN PUT ;
      RETURN;

      REPORT_HEAD:
         PUT @1 "Exceptions Report for Data Set &DATASET" /
             "Using Validation Data File &VALID_FILE"  //
            @1  "Patient ID"
            @18 "Variable"
            @29 "Value"
            @39 "Reason" /
            @1  60*"-";
   RUN;

   PROC DATASETS LIBRARY=WORK;
      DELETE PAT VALID;
   RUN;
   QUIT;

%MEND VALID_CHAR;
```

To test this macro, execute the following macro call:

```
%VALID_CHAR(PATNO,TEST_CHAR,C:\CLEANING\VALID_C.TXT)
```

The resulting listing is identical to the listing from the nonmacro version of this program.

It is possible to extend this program (or macro) to handle more complicated validation rules. For example, you may want to check for valid character values in a range (for example, AAA – ZZZ). This is demonstrated in the next section.

Extending the Validation Macro to Include Valid Character Ranges

Besides listing discrete character values for each of the character variables to be tested, you might want to indicate a range of possible values such as 'A' to 'E'. The macro that follows does just that. It allows the user to include discrete character values as well as ranges in the validation data set. For example, to check for the values 'A', 'B', 'AB', 'C', 'DDD' to 'FFF', and 'X' to 'Z', for a variable called CODE, you would enter the line:

```
CODE A B AB C DDD-FFF X-Z    Y
```

The program needs to search the string (VALUES_LIST) for dashes and treat the strings directly before and directly after it, as the beginning and ending values for a range. Luckily for us, SAS has a good selection of character functions. This macro is described (but not in too much detail) after Program 9-9.

Program 9-9   Writing a Macro to Check for Discrete Character Values and Character Ranges

```
*-------------------------------------------------------------------*
| Program Name: RANGE.SAS  in C:\CLEANING                           |
| Purpose: This macro takes an ID variable, a SAS data set to be    |
|          validated, and a validation data file and checks for     |
|          discrete character values or character ranges for        |
|          valid data, and prints an exception report.              |
| Arguments: ID          - ID variable name                         |
|            DATASET      - SAS data set to be validated            |
|            VALID_FILE - Validation data file containing variable |
|                          names, discrete valid character values   |
|                          and/or ranges, and a missing value flag. |
|                          'Y' means missing values are OK.         |
| Example: %RANGE(PATNO,TEST_CHAR,C:\CLEANING\VALID_RANGE.TXT)      |
*-------------------------------------------------------------------*;
%MACRO RANGE(ID,           /* ID variable              */
             DATASET,      /* Data set to be validated */
             VALID_FILE,   /* Validation data file     */
             );

   PROC FORMAT;
      INVALUE $MISS(UPCASE DEFAULT=1) 'Y' = 'Y'
                                      OTHER = 'N';
   RUN;

   DATA C_VALID;
      LENGTH VARNAME $ 32 VALUES_LIST $ 200 MISS_OK $ 1 WORD $ 17;
      INFILE "&VALID_FILE" MISSOVER;
      INPUT VARNAME VALUES_LIST & $200. MISS_OK : $MISS.;
      ***Separate VALUES_LIST into individual values and ranges;

      ***Array to store up to 10 ranges.  The first dimension of
         the array tells which range it is, the second dimension
         takes on the value 1 for the lower range and 2 for the
         upper range.  You may want to increase the length for
         each of the ranges to a larger number. ;

      ARRAY RANGES[10,2] $ 8 R1-R20;    ❶

      ***Compute the number of ranges in the string;
      N_OF_RANGES = LENGTH(VALUES_LIST) -
                 LENGTH(COMPRESS(VALUES_LIST,"-"));    ❷
```

```
   ***Break list into "words";
   N_RANGE = 0;
   DO I = 1 TO 200 UNTIL (WORD = " ");   ❸
      WORD = SCAN(VALUES_LIST,I," ");
      IF INDEX(WORD,'-') NE 0 THEN DO;   ❹
         ***Range found, scan again to get lower and
         upper values;
         N_RANGE + 1;
         RANGES[N_RANGE,1] = SCAN(WORD,1,"-");
         RANGES[N_RANGE,2] = SCAN(WORD,2,"-");
      END;
   END;

   ***When all finished finding ranges, substitute spaces for dashes;
   VALUES_LIST = TRANSLATE(VALUES_LIST," ","-");

   KEEP VALUES_LIST R1-R20 VARNAME N_OF_RANGES MISS_OK ;
RUN;

***Restructure TEST_CHAR;
DATA PAT;
   SET TEST_CHAR;
   ARRAY CHARS[*] _CHARACTER_;
   LENGTH VARNAME $ 32;
   N_CHARS = DIM(CHARS);
   DO I = 1 TO N_CHARS;
      CALL VNAME(CHARS[I],VARNAME);
      VARNAME = UPCASE(VARNAME);
      VALUE = CHARS[I];
      OUTPUT;
   END;
   KEEP PATNO VARNAME VALUE;
RUN;

PROC SORT DATA=PAT;
   BY VARNAME;
RUN;

PROC SORT DATA=C_VALID;
   BY VARNAME;
RUN;
```

```
DATA VERIFY;
   ARRAY RANGES[10,2] $ 8 R1-R20;
   MERGE PAT(IN=IN_PAT) C_VALID(IN=IN_C_VALID);
   BY VARNAME;
   IF (IN_PAT AND IN_C_VALID);

   LENGTH TOKEN $ 8;

   ***Obviously bad values;
   IF (VERIFY (VALUE,VALUES_LIST) NE 0 AND N_OF_RANGES = 0)     OR
      VALUE = ' ' AND MISS_OK NE 'Y' THEN DO;
      OUTPUT;
      RETURN;
   END;

   ***Checking for discrete values;
   FLAG = 0;   /* Flag incremented if a discrete match found */
   DO I = 1 TO 99;
      TOKEN = SCAN(VALUES_LIST,I);
      IF VALUE = TOKEN THEN FLAG + 1;
      IF TOKEN = ' '  OR FLAG > 0 THEN LEAVE;
   END;

   ***Checking for ranges;
   ***R_FLAG incremented if in one of the ranges;   ❺
   R_FLAG = 0;
   ***Lower and upper range values already checked above;
   IF N_OF_RANGES > 0 THEN DO I = 1 TO N_OF_RANGES;
   IF VALUE > RANGES[I,1] AND VALUE < RANGES[I,2] THEN DO;
         R_FLAG + 1;
         LEAVE;
      END;
   END;

   IF FLAG = 0 AND R_FLAG = 0 THEN OUTPUT;

   DROP I TOKEN;
RUN;

PROC SORT DATA=VERIFY;
   BY PATNO VARNAME;
RUN;
```

```
***Reporting section;
OPTIONS NODATE NONUMBER;
TITLE;

DATA _NULL_;
   FILE PRINT HEADER = REPORT_HEAD;
   SET VERIFY;
   BY PATNO;

   IF VALUE = ' ' THEN PUT
      @1   PATNO
      @18 VARNAME
      @39 "Missing";

   ELSE   PUT
      @1   PATNO
      @18 VARNAME
      @29 VALUE
      @39 "Not Valid";

   IF LAST.PATNO THEN PUT ;
   RETURN;

   REPORT_HEAD:
      PUT @1 "Exceptions Report for Data Set TEST_CHAR" /
            "Using Validation Data Set VALID_C"  //
         @1   "Patient ID"
         @18 "Variable"
         @29 "Value"
         @39 "Reason" /
         @1  60*"-";
RUN;

PROC DATASETS LIBRARY=WORK NOLIST;
   DELETE PAT   C_VALID;
RUN;
QUIT;

%MEND RANGE;
```

The first part of this macro creates the validation data set from the raw data file, using the user-defined informat $MISS as before. Using a two-dimensional array, you can hold up to 10 separate ranges ❶. The first dimension tells you which range you are working with, and the second dimension tells whether the value is the lower (value of 1) or upper (value of 2) end of the range. A quick and easy way to count the number of ranges is to count the number of dashes in the VALID_LIST string. This is accomplished by a useful trick: take the length of the original string and subtract the length of the string after you have removed (COMPRESS function) the dashes ❷.

For each of the ranges, you have to extract the lower and upper value. This is accomplished by the statements starting with the DO loop ❸. First, scan the entire VALUES_LIST and break it into "words." If a word contains a dash ❹, you know that you have located a range, rather than a discrete character value. Use the SCAN function again, this time using a dash as the delimiter. The portion of the word before the dash is the lower range and is stored in the array element RANGES[N_RANGE,1]. The portion after the dash is the upper range and is stored in the array element RANGES[N_RANGE,2].

Restructuring the data set to be validated is the same as before. This time, a section was added to check for ranges ❺. The remainder of the program is identical to Program 9-8.

To test this macro, call it as follows:

```
%RANGE(PATNO,TEST_CHAR,C:\CLEANING\VALID_RANGE.TXT)
```

where the contents of the raw data file VALID_RANGE.TXT is

```
CODE A B AB C DDD-FFF X-Z    Y
GENDER M F   N
AE 0 1    Y
```

The report generated by this macro is shown next.

```
Exceptions Report For Data Set TEST_CHAR
Using Validation Data Set C:\CLEANING\VALID_RANGE.TXT

Patient ID      Variable   Value    Reason
-----------------------------------------------------------
003             CODE       BA       Not Valid

004             GENDER              Missing

005             AE         Z        Not Valid
005             GENDER     Y        Not Valid

006             CODE       AC       Not Valid
```

Combining Numeric and Character Validity Checks in a Single Macro with a Single Validation Data Set

In this section, the two previous macros are combined, one to test for valid numeric values and the other to check for discrete character values or character ranges. The validation data file includes information on both the numeric and character variables, in any order. For the numeric variables, include the variable name (in uppercase or lowercase), the minimum and maximum values, and a 'Y' if missing values are okay (not to be flagged as errors). For the character variables, include the variable name, a list of discrete character values and/or ranges of character values, and a missing value indicator. Be sure to insert at least two spaces between the values list and the 'Y' (or 'N'). The discrete character values and ranges can be in any order.

For example, to write a validation file for the following three rules:

- a variable called CODE, with acceptable values of 'A', 'B', 'K' - 'P', and 'FF' - 'HH', where missing values are okay
- a variable called X, with a minimum and maximum value of 10 and 20, where missing values are not okay (to be considered as errors)
- a variable called CHOICE, with acceptable values of 'W', 'B', and 'A', where missing values are not okay (to be considered as errors)

you enter the following three data lines into your validation file:

```
CODE A B K-P FF-HH  Y
X 10 20
CHOICE W B A  N
```

Note: You can include an 'N' as a missing value indicator or leave it out. Any nonmissing character other than a 'Y' is converted to an 'N' by the program. Also, remember that you need at least two spaces between the list of character values and the missing value flag.

Much of the programming is similar to the programs above, so the logic doesn't need to be explained. However, following the listing, the parts that are new are explained.

Program 9-10 Creating a Macro to Validate both Numeric and Character Data, Including Character Ranges, with a Single Validation Data File

```
*-------------------------------------------------------------------*
| Program Name: VALID_ALL.SAS  in C:\CLEANING                       |
| Purpose: This macro takes an ID variable, a SAS data set to be    |
|          validated, and a validation data file, and checks for    |
|          discrete character values or character ranges for        |
|          character variables and valid ranges for numeric data    |
|          and prints an exception report.                          |
| Arguments: ID         - ID variable name                          |
|            DATASET    - SAS data set to be validated              |
|            VALID_FILE - Validation data file containing variable  |
|                         names, discrete values and/or ranges for  |
|                         character variables, minimum and maximum  |
|                         values for numeric variables, and a       |
|                         missing value flag 'Y' means missing      |
|                         values are okay.                          |
| Example: %VALID_ALL(PATNO,CLEAN.PATIENTS,C:\CLEANING\VALID_ALL.TXT)|
*-------------------------------------------------------------------*;
%MACRO VALID_ALL(ID,         /* ID variable              */
                 DATASET,    /* Data set to be validated */
                 VALID_FILE /* Validation data file      */
                 );

   ***Get a list of variable names and type (numeric or character);
   PROC CONTENTS NOPRINT DATA=&DATASET
                 OUT=NAMETYPE(KEEP=NAME TYPE);   ❶
   RUN;

   ***Find number of observations in data set TYPE and assign
      to a macro variable;
   %LET DSID = %SYSFUNC(OPEN(NAMETYPE));   ❷
   %LET NUM_OBS = %SYSFUNC(ATTRN(&DSID,NOBS));
   %LET RC = %SYSFUNC(CLOSE(&DSID));

   ***Place the variable names and types in a single observation,
      using NAMES1-NAMESn and TYPE_VAR1-TYPE_VARn to hold the
      variable names and types, respectively;
```

```
DATA XTYPE;  ❸
   SET NAMETYPE END=LAST;
   NAME = UPCASE(NAME);
   ARRAY NAMES[&NUM_OBS] $ 32;
   ARRAY TYPE_VAR[&NUM_OBS];
   RETAIN  NAMES1-NAMES&NUM_OBS  TYPE_VAR1-TYPE_VAR&NUM_OBS;
   NAMES[_N_] = NAME;
   TYPE_VAR[_N_] = TYPE;
   IF LAST THEN OUTPUT;
   KEEP NAMES1-NAMES&NUM_OBS TYPE_VAR1-TYPE_VAR&NUM_OBS;
RUN;


***Turn the validation data file into a SAS data set;
PROC FORMAT;
   INVALUE $MISS(UPCASE DEFAULT=1) 'Y' = 'Y'
                                   OTHER = 'N';
RUN;


***Need to distinguish lines with numeric ranges from ones
   with character values and ranges.  Use the variable TYPE
   from the one observation data set (XTYPE) above;

DATA VALID;
   ARRAY NAMES[&NUM_OBS] $ 32;
   ARRAY TYPE_VAR[&NUM_OBS];
   LENGTH VARNAME $ 32 VALUES_LIST $ 200 MISS_OKAY $ 1 WORD $ 17;
   INFILE "&VALID_FILE" MISSOVER;
   IF _N_ = 1 THEN SET XTYPE;  ❹
   INPUT VARNAME @;  ❺
   VARNAME = UPCASE(VARNAME);

   ***Find VARNAME in NAMES array and determine TYPE;
   DO I = 1 TO &NUM_OBS;
      IF VARNAME = NAMES[I] THEN DO;
         IF TYPE_VAR[I] = 1 THEN DO;
            INPUT MIN MAX MISS_OKAY : $MISS.;
            TYPE = 'N';
         END;

      ELSE IF TYPE_VAR[I] = 2 THEN DO;
         INPUT VALUES_LIST & $200. MISS_OKAY : $MISS.;
         ***Separate VALUES_LIST into individual
            values and ranges;
         ARRAY RANGES[10,2] $ 8 R1-R20;
         N_OF_RANGES = LENGTH(VALUES_LIST) -
                     LENGTH(COMPRESS(VALUES_LIST,"-"));
```

```
              ***Break list into "words";
              N_RANGE = 0;
              DO I = 1 TO 200 UNTIL (WORD = " ");
                 WORD = SCAN(VALUES_LIST,I," ");
                 IF INDEX(WORD,'-') NE 0 THEN DO;
                    ***Range found, scan again to get lower and
                    upper values;
                    N_RANGE + 1;
                    RANGES[N_RANGE,1] = SCAN(WORD,1,"-");
                    RANGES[N_RANGE,2] = SCAN(WORD,2,"-");
                 END;
              END;

              TYPE = 'C';
          END;

          OUTPUT;
          LEAVE;
       END;
   END;
   KEEP VARNAME MIN MAX MISS_OKAY VALUES_LIST TYPE
          R1-R20  N_OF_RANGES;
RUN;

***Restructure data set to be validated.  Need separate
   variable to hold numeric and character values;
DATA PAT;
   SET &DATASET;
   ARRAY CHARS[*] _CHARACTER_;      ❻
   ARRAY NUMS[*] _NUMERIC_;         ❼
   LENGTH VARNAME $ 32;
   N_CHARS = DIM(CHARS);
   N_NUMS = DIM(NUMS);

   DO I = 1 TO N_CHARS;             ❽
      CALL VNAME(CHARS[I],VARNAME);
      VARNAME = UPCASE(VARNAME);
      C_VALUE = CHARS[I];           ❾
      OUTPUT;
   END;
```

```
   DO I = 1 TO N_NUMS;  ❿
      CALL VNAME(NUMS[I],VARNAME);
      VARNAME = UPCASE(VARNAME);
      N_VALUE = NUMS[I];  ⓫
      C_VALUE = " ";
      OUTPUT;
   END;

   KEEP PATNO VARNAME C_VALUE N_VALUE;
RUN;

PROC SORT DATA=PAT;
   BY VARNAME;
RUN;

PROC SORT DATA=VALID;
   BY VARNAME;
RUN;

DATA VERIFY;
   ARRAY RANGES[10,2] $ 8 R1-R20;
   MERGE PAT(IN=IN_PAT) VALID(IN=IN_VALID);
   BY VARNAME;
   IF NOT(IN_PAT AND IN_VALID) THEN DELETE;

   LENGTH TOKEN $ 8;

   ***Character variable section;
   IF TYPE = 'C' THEN DO;

      ***Obviously bad values;
      IF (VERIFY (C_VALUE,VALUES_LIST) NE 0 AND N_OF_RANGES = 0) OR
         C_VALUE = ' ' AND MISS_OKAY = 'N' THEN DO;
         OUTPUT;
         RETURN;
      END;

      ***Checking for discrete values;
      FLAG = 0;
      DO I = 1 TO 99;
         TOKEN = SCAN(VALUES_LIST,I);
         IF C_VALUE = TOKEN THEN FLAG + 1;
         IF TOKEN = ' '  OR FLAG > 0 THEN LEAVE;
      END;
```

```
   ***Checking for ranges;
   R_FLAG = 0;
   IF N_OF_RANGES > 0 THEN DO I = 1 TO N_OF_RANGES;
      IF C_VALUE > RANGES[I,1] AND
         C_VALUE < RANGES[I,2] THEN DO;
         R_FLAG + 1;
         LEAVE;
      END;
   END;

   IF FLAG = 0 AND R_FLAG = 0 THEN OUTPUT;

END;
***End of character section;

***Numeric variable section;
IF TYPE = 'N' THEN DO;
   IF   (N_VALUE LT MIN OR N_VALUE GT MAX)  AND
        NOT(N_VALUE = . AND MISS_OKAY EQ 'Y') THEN OUTPUT;
END;
***End of numeric section;

DROP VALUES_LIST TOKEN I FLAG;
RUN;


PROC SORT DATA=VERIFY;
   BY PATNO VARNAME;
RUN;


***Reporting section;
OPTIONS NODATE NONUMBER;
TITLE;


DATA _NULL_;
   FILE PRINT HEADER = REPORT_HEAD;
   SET VERIFY;
   BY PATNO;

   ***Numeric variables;
   IF TYPE = 'N' THEN DO;
   IF N_VALUE = . THEN PUT
      @1   PATNO
      @18  VARNAME
      @39  "Missing";
```

```
ELSE IF N_VALUE GT . AND N_VALUE LT MIN THEN PUT
    @1   PATNO
    @18 VARNAME
    @29 N_VALUE
    @39 "Below Minimum ("  MIN  +(-1) ")";

ELSE IF N_VALUE GT MAX THEN PUT
    @1   PATNO
    @18 VARNAME
    @29 N_VALUE
    @39 "Above Maximum ("  MAX  +(-1) ")";

IF LAST.PATNO THEN PUT ;
END;
***End of numeric report;

***Character report;
IF TYPE = 'C' THEN DO;
IF C_VALUE = ' ' THEN PUT
    @1   PATNO
    @18 VARNAME
    @39 "Missing";

ELSE  PUT
    @1   PATNO
    @18 VARNAME
    @29 C_VALUE
    @39 "Not Valid";

IF LAST.PATNO THEN PUT ;
END;
***End of character report;

RETURN;

REPORT_HEAD:
    PUT @1 "Exceptions Report for Data Set &DATASET" /
            "Using Validation Data File &VALID_FILE"  //
        @1  "Patient ID"
        @18 "Variable"
        @29 "Value"
        @39 "Reason" /
        @1  60*"-";
RUN;
```

```
   ***Cleanup temporary data sets;
   PROC DATASETS LIBRARY=WORK NOLIST;
      DELETE PAT;
      DELETE VERIFY;
      DELETE NAMETYPE;
   RUN;
   QUIT;

RUN;


%MEND VALID_ALL;
```

At the beginning of this macro, PROC CONTENTS ❶ is used to create an output data set (NAMETYPE), which contains the variable name and type (1=numeric, 2=character). Next, the macro function %SYSFUNC is called three times, to place the number of observations in the data set NAMETYPE (the output data set from PROC CONTENTS) into the macro variable NUM_OBS ❷. (See *SAS Macro Programming Made Easy*, p. 128, by Michele Burlew for more details on %SYSFUNC).

Starting with line ❸, you create a data set that has one observation with variables NAMES1-NAMES*n* and TYPE_VAR1-TYPE_VAR*n*, where *n* is the number of observations in the NAMETYPE data set. This data set will be used later to look up a variable name and determine its type.

In order to read the validation data file and create the validation data set, you need to decide whether you are reading a line corresponding to a numeric variable (the variable name, minimum and maximum values, and a missing flag) or one corresponding to a character variable (the variable name, the list of valid character values or ranges, and a missing flag). You do this by bringing in the NAME and TYPE_VAR variables from the data set XTYPE into the PDV ❹. Next, read in the variable name from the validation file ❺. Then loop through each of the variable names in the NAMES*n* variables, looking for a match. If you find a match, you know the corresponding TYPE_VAR*n* variable holds the variable type. Depending on this value, you read either the information for numeric or character variables. You also set the variable TYPE to either an 'N' or a 'C'.

When you restructure the data set to be validated, you set up two arrays, one for the character variables ❻ and one for the numeric variables ❼. (Remember, you cannot mix character and numeric variables in the same array.) This also makes it easy to know whether a given variable is character or numeric.

You have two DO loops, one for the character variables ❽ and one for the numeric variables ❿. Because you need to distinguish between the two, use two different variable names, C_VALUE and N_VALUE (❾ and ⓫, respectively) to store the value of the character and numeric variables. Finally, use the same logic, as used in Programs 9-8 and 9-9, to see if a numeric or a character value is invalid.

The reporting sections are also the same as before. Here, the TYPE variable also directs which type of report line to write, depending on whether it has a value of 'C' or 'N'.

Before demonstrating this macro, I would advise caution before using it in a production environment. (As they say, "your mileage may vary" or in this case, the car may not run at all!) It is possible that there will be situations where this program fails or produces erroneous error reports. As with just about any SAS program or macro, careful testing on a sample data set is essential.

To demonstrate this macro, let's create a validation data file to validate the data in the original data set CLEAN.PATIENTS. The rules are similar to the ones used in previous chapters. They are

Patient number (PATNO) is to be a three-digit numeral. Missing values are not allowed for this variable. (See the Important Note in Chapter 1, page 14, concerning the range 001-999.)

Gender (GENDER) has valid values of 'M' and 'F'. Missing values are permitted.

Heart rate (HR) must be within the range 40 to 100. Missing values are not valid.

Systolic blood pressure (SBP) must be within the range 80 to 200. Missing values are permitted.

Diastolic blood pressure (DBP) must be within the range 60 to 120. Missing values are permitted.

Diagnosis code (DX) is a one-, two-, or three-digit numeral. Missing values are permitted. (The same caveat pertains to this as in the first rule above.)

Adverse events (AE) has valid codes of '0' and '1'. Missing values are permitted.

The raw data file C:\CLEANING\VALID_ALL.TXT, which follows these seven rules, is shown next.

```
patno 001-999      N
gender M F         Y
hr 40 100          N
sbp 80 200         Y
dbp 60 120         y
dx 001-999         Y
ae 0 1             y
```

First, as we mentioned in Chapter 1, the character range 001 to 999 may contain such values as '0A9', depending on your operating environment. If that is the case, you should run an additional check on the patient numbers using a short DATA step. The same is true for DX codes. To run the validation program on the data set CLEAN.PATIENTS, using the data file C:\CLEANING\VALID_ALL.TXT, submit the following statement:

```
%VALID_ALL(PATNO,CLEAN.PATIENTS,C:\CLEANING\VALID_ALL.TXT)
```

which results in the following error report:

```
Exceptions Report For Data Set clean.patients
Using Validation Data File c:\cleaning\VALID_ALL.TXT

Patient ID      Variable   Value    Reason
-----------------------------------------------------------
                PATNO               Missing

002             DX         X        Not Valid
002             DX         X        Not Valid

003             GENDER     X        Not Valid

004             AE         A        Not Valid
004             HR         101      Above Maximum (100)

008             HR         210      Above Maximum (100)

009             DBP        180      Above Maximum (120)
009             SBP        240      Above Maximum (200)

010             GENDER     f        Not Valid
010             HR                  Missing
010             SBP        40       Below Minimum (80)

011             DBP        20       Below Minimum (60)
011             SBP        300      Above Maximum (200)

013             GENDER     2        Not Valid

014             HR         22       Below Minimum (40)

                                                 Continued
```

```
017                 HR          208 ·    Above Maximum (100)

020                 DBP         8        Below Minimum (60)
020                 HR          10       Below Minimum (40)
020                 SBP         20       Below Minimum (80)

023                 GENDER      f        Not Valid
023                 HR          22       Below Minimum (40)
023                 SBP         34       Below Minimum (80)

027                 HR                   Missing

029                 HR                   Missing

321                 DBP         200      Above Maximum (120)
321                 HR          900      Above Maximum (100)
321                 SBP         400      Above Maximum (200)

XX5                 PATNO       XX5      Not Valid
```

That's about as far as this book will take the concept of validation data sets. You can see that, at some point, the complexity of the code reaches the point of diminishing returns. You may very well want to stick to some of the more straightforward techniques described in the earlier chapters. Hopefully, these programs will help you generate some novel ideas of your own.

Introducing SAS Integrity Constraints (Versions 7 and Later)

Starting with Version 7 of SAS software, a feature called integrity constraints was implemented. Integrity constraints are rules that are stored with a SAS data set that can restrict data values accepted into the data set when new data is added with PROC APPEND, DATA step MODIFY, and SQL insert, delete, or update. These constraints are preserved when the data set is copied using PROC COPY, CPORT, or CIMPORT or is sorted with PROC SORT.

Briefly, there are two types of integrity constraints. One type, called general integrity constraints, allows you to restrict data values that are added to a SAS data set. You can specify specific values for a variable, a range of values, or a requirement that values of a variable be unique. The other type, called referential integrity constraints, allows you to link data values between files. For example, you may have a list of valid patient numbers in a demographic file. You could restrict any patient numbers added to another file to those that exist in the demographic file.

Integrity constraints can keep a data set "pure" by rejecting any observations that violate one or more of the constraints. This feature probably has more utility in a data warehouse than in a clinical data application. The reason for this is that when you attempt to append data to an existing data set that contains integrity constraints, those observations that are valid are added, and those observations that violate one or more of the integrity constraints are rejected. The SAS Log does not give you information about which observations were accepted or rejected — you will have to determine that on your own. (This may change with later releases of SAS software.)

Integrity constraints can be created with PROC SQL statements or PROC DATASETS. To demonstrate how you could use integrity constraints to prevent data values being added to an existing SAS data set, let's first create a small data set called IC_DEMO by running the following program:

```
DATA IC_DEMO;
    INPUT PATNO : $3. GENDER : $1.  HR    SBP    DBP;
DATALINES;
001 M  88 140   80
002 F  84 120   78
003 M  58 112   74
004 F  . 200 120
007 M  88 148 102
015 F  82 148   88
;
```

A listing of this data set is shown next.

```
Listing of IC_DEMO

Obs    PATNO    GENDER    HR    SBP    DBP

1       001       M       88    140    80
2       002       F       84    120    78
3       003       M       58    112    74
4       004       F       .     200   120
5       007       M       88    148   102
6       015       F       82    148    88
```

The next program will add integrity constraints to the IC_DEMO data set.   The constraints are

- GENDER must be 'F' or 'M'.

- HR (heart rate) must be between 40 and 100. Missing values are allowed.

- SBP (systolic blood pressure) must be between 80 and 200. Missing values are allowed.

- DBP (diastolic blood pressure) must be between 60 and 140. Missing values are not allowed.

- PATNO (patient number) must be unique.

Here are the PROC DATASETS statements.

```
PROC DATASETS LIBRARY=WORK NOLIST;
   MODIFY IC_DEMO;
      IC CREATE GEN_CHK = CHECK
         (WHERE=(GENDER IN('F','M')));

      IC CREATE HR_CHK = CHECK
         (WHERE=( HR BETWEEN 40 AND 100 OR HR = .));

      IC CREATE SBP_CHK = CHECK
         (WHERE=(SBP BETWEEN 80 AND 200 OR SBP IS NULL));

      IC CREATE DBP_CHK = CHECK
         (WHERE=(DBP BETWEEN 60 AND 140));

      IC CREATE ID_CHK = UNIQUE(PATNO);
QUIT;
```

Running PROC CONTENTS will display the usual data set information as well as the integrity constraints.   The PROC CONTENTS statements and the resulting output (edited) are shown next.

```
PROC CONTENTS DATA=IC_DEMO;
   TITLE "Output from PROC CONTENTS";
RUN;
```

```
Output from PROC CONTENTS

The CONTENTS Procedure

Data Set Name: WORK.IC_DEMO        Observations:              6
Member Type:   DATA                Variables:                 5
Engine:        V7                  Indexes:                   1
Created:       14:41 Tuesday,      Integrity Constraints:  5
               May 4, 1999
Last Modified: 14:42 Tuesday,      Observation Length:       32
               May 4, 1999
Protection:                        Deleted Observations:     0
Data Set Type:                     Compressed:               NO
Label:                             Sorted:                   YES

        -----Alphabetic List of Variables and Attributes-----

  #    Variable    Type    Len   Pos    Label
  ------------------------------------------------------------------
  5    DBP         Num      8    16     Diastolic Blood Pressure
  2    GENDER      Char     1    27     Gender
  3    HR          Num      8     0     Heart Rate
  1    PATNO       Char     3    24     Patient Number
  4    SBP         Num      8     8     Systolic Blood Pressure


          -----Alphabetic List of Integrity Constraints-----

    Integrity                   WHERE
  # Constraint Type   Variables Clause
  ------------------------------------------------------------------
  1 dbp_chk    Check            ((DBP>=60 and DBP<=140))
  2 gen_chk    Check            GENDER in ('F', 'M')
  3 hr_chk     Check            (HR=. or (HR>=40 and HR<=100))
  4 id_chk     Unique PATNO
  5 sbp_chk    Check            ((SBP>=80 and SBP<=200))
                                or (SBP is null)
```

Notice that each of the WHERE clauses that created the integrity constraints are listed in the output from PROC CONTENTS.

What happens when you try to append data that violates one or more of the integrity constraints? The short DATA step shown next creates a data set (NEW) in which the second observation (PATNO=567) violates the integrity constraints for HR. Let's run PROC APPEND next, in an attempt to append the new data to the IC_DEMO data set.

```
DATA NEW;
    INPUT PATNO : $3. GENDER : $1. HR SBP DBP;
DATALINES;
456 M 66 98 72
567 F 150 130 80
;
PROC APPEND BASE=IC_DEMO DATA=NEW;
RUN;
```

The SAS Log (shown next) shows that one observation had a data value that violated the heart rate integrity constraint (HR_CHK).

```
38    ;
39    PROC APPEND BASE=IC_DEMO DATA=NEW;
40    RUN;

NOTE: Appending WORK.NEW to WORK.IC_DEMO.
WARNING: Data value(s) do not comply with integrity constraint
HR_CHK for file IC_DEMO, 1 observations rejected.
NOTE: 1 observations added.
NOTE: The data set WORK.IC_DEMO has 7 observations and 5 variables.
NOTE: PROCEDURE APPEND used:
      real time            0.33 seconds
```

As mentioned in the beginning of this section, the SAS Log does not tell you which observation was added and which was rejected. It does ensure that only observations meeting the integrity constraints can be appended to the original data set.

Let's test the UNIQUE property of the PATNO variable. The next DATA step creates a three-observation data set (NEW2) in which the first two observations contain patient numbers that are already in the IC_DEMO data set. Let's see what happens when you attempt to append it to the IC_DEMO data set.

```
DATA NEW2;
    INPUT PATNO : $3. GENDER : $1. HR SBP DBP;
DATALINES;
003 M 66 98 72
015 F 80 130 80
777 F 70 110 70
;
PROC APPEND BASE=IC_DEMO DATA=NEW2;
RUN;
```

Here is the resulting SAS Log.

```
47   ;
48   PROC APPEND BASE=IC_DEMO DATA=NEW2;
49   RUN;

NOTE: Appending WORK.NEW2 to WORK.IC_DEMO.
WARNING: Data value(s) do not comply with integrity constraint
ID_CHK for file IC_DEMO, 2 observations rejected.
NOTE: 1 observations added.
NOTE: The data set WORK.IC_DEMO has 8 observations and 5 variables.
NOTE: PROCEDURE APPEND used:
      real time            0.04 seconds
```

This section gives only a brief glimpse of integrity constraints. Two sources of additional information about integrity constraints are the SAS Online Documentation for Version 7 (or later) and the SAS Web site:

www.sas.com/techsup/download/technote/ts594.html

Appendix

Listing of Raw Data Files and SAS Programs

Description of the Raw Data File PATIENTS.TXT

The raw data file PATIENTS.TXT contains both character and numeric variables from a typical clinical trial. A number of data errors were included in the file so that you can test the data cleaning programs that are developed in this text. The programs in this book assume that the file PATIENTS.TXT is located in a directory (folder) called C:\CLEANING. This is the directory that is used throughout this text as the location for data files, SAS data sets, SAS programs, and SAS macros. You should be able to modify the INFILE and LIBNAME statements to fit your own operating environment.

Layout for the Data File PATIENTS.TXT

Variable Name	Description	Starting Column	Length	Variable Type	Valid Values
PATNO	Patient Number	1	3	Character	Numerals only
GENDER	Gender	4	1	Character	'M' or 'F'
VISIT	Visit Date	5	10	MMDDYY10.	Any valid date
HR	Heart Rate	15	3	Numeric	Between 40 and 100
SBP	Systolic Blood Pressure	18	3	Numeric	Between 80 and 200
DBP	Diastolic Blood Pressure	21	3	Numeric	Between 60 and 120
DX	Diagnosis Code	24	3	Character	1 to 3 digit numeral
AE	Adverse Event	27	1	Character	'0' or '1'

Listing of Raw Data File PATIENTS.TXT

```
12345678901234567890012345 (ruler)
-----------------------------------
001M11/11/1998 88140 80   10
002F11/13/1998 84120 78   X0
003X10/21/1998 68190100   31
004F01/01/1999101200120   5A
XX5M05/07/1998 68120 80   10
006 06/15/1999 72102 68   61
007M08/32/1998 88148102    0
   M11/11/1998 90190100    0
008F08/08/1998210         70
009M09/25/1999 86240180   41
010f10/19/1999     40120   10
011M13/13/1998 68300 20   41
012M10/12/98    60122 74    0
013208/23/1999 74108 64    1
014M02/02/1999 22130 90    1
002F11/13/1998 84120 78   X0
003M11/12/1999 58112 74    0
015F          82148 88   31
017F04/05/1999208     84   20
019M06/07/1999 58118 70    0
123M15/12/1999 60         10
321F          900400200   51
020F99/99/9999 10 20  8    0
022M10/10/1999 48114 82   21
023f12/31/1998 22 34 78    0
024F11/09/199876 120 80   10
025M01/01/1999 74102 68   51
027FNOTAVAIL  NA 166106   70
028F03/28/1998 66150 90   30
029M05/15/1998            41
006F07/07/1999 82148 84   10
```

Program to Create the SAS Data Set PATIENTS

```
*-----------------------------------------------------------------*
| PROGRAM NAME: PATIENTS.SAS IN C:\CLEANING                       |
| PURPOSE: TO CREATE A SAS DATA SET CALLED PATIENTS               |
| DATE: MAY 29, 1998                                              |
*-----------------------------------------------------------------*;
LIBNAME CLEAN "C:\CLEANING";

DATA CLEAN.PATIENTS;
   INFILE "C:\CLEANING\PATIENTS.TXT" PAD;
   INPUT @1  PATNO    $3.
         @4  GENDER   $1.
         @5  VISIT    MMDDYY10.
         @15 HR       3.
         @18 SBP      3.
         @21 DBP      3.
         @24 DX       $3.
         @27 AE       $1.;

   LABEL PATNO   = "Patient Number"
         GENDER  = "Gender"
         VISIT   = "Visit Date"
         HR      = "Heart Rate"
         SBP     = "Systolic Blood Pressure"
         DBP     = "Diastolic Blood Pressure"
         DX      = "Diagnosis Code"
         AE      = "Adverse Event?";

   FORMAT VISIT MMDDYY10.;

RUN;
```

Listing of Raw Data File PATIENTS2.TXT

```
Listing of the File PATIENTS2.TXT

          1         2
1234567890123456789012345 (ruler)
---------------------------------
00106/12/1998 80130 80
00106/15/1998 78128 78
00201/01/1999 48102 66
00201/10/1999 70112 82
00202/09/1999 74118 78
00310/21/1998 68120 70
00403/12/1998 70102 66
00403/13/1998 70106 68
00504/14/1998 72118 74
00504/14/1998 74120 80
00611/11/1998100180110
00709/01/1998 68138100
00710/01/1998 68140 98
```

Program to Create the SAS Data Set PATIENTS2

```
LIBNAME CLEAN "C:\CLEANING";

DATA CLEAN.PATIENTS2;
    INFILE "C:\CLEANING\PATIENTS2.TXT" PAD;
    INPUT @1  PATNO $3.
          @4  VISIT MMDDYY10.
          @14 HR      3.
          @17 SBP     3.
          @20 DBP     3.;
    FORMAT VISIT MMDDYY10.;
RUN;
```

Program to Create the SAS Data Set AE (Adverse Events)

```
LIBNAME CLEAN "C:\CLEANING";

DATA CLEAN.AE;
   INPUT @1  PATNO    $3.
         @4   DATE_AE MMDDYY10.
         @14 A_EVENT $1.;
   LABEL PATNO   = 'Patient ID'
         DATE_AE = 'Date of AE'
         A_EVENT = 'Adverse Event';
   FORMAT DATE_AE MMDDYY10.;
DATALINES;
00111/21/1998W
00112/13/1998Y
00311/18/1998X
00409/18/1998O
00409/19/1998P
01110/10/1998X
01309/25/1998W
00912/25/1998X
02210/01/1998W
02502/09/1999X
;
```

Program to Create the SAS Data Set LAB_TEST

```
LIBNAME CLEAN "C:\CLEANING";

DATA CLEAN.LAB_TEST;
   INPUT @1   PATNO     $3.
         @4   LAB_DATE DATE9.
         @13  WBC       5.
         @18  RBC       4.;
   LABEL  PATNO     = 'Patient ID'
          LAB_DATE = 'Date of Lab Test'
          WBC      = 'White Blood Cell Count'
          RBC      = 'Red Blood Cell Count';
   FORMAT LAB_DATE MMDDYY10.;
DATALINES;
00115NOV1998 90005.45
00319NOV1998 95005.44
00721OCT1998 82005.23
00422DEC1998110005.55
02501JAN1999 82345.02
02210OCT1998 80005.00
;
```

Index

Books Available from SAS Press

support.sas.com/pubs

SAS® System for Elementary Statistical Analysis,
Second Edition
by **Sandra D. Schlotzhauer**
and **Ramon C. Littell**. . . . . Order No. A55172

SAS® System for Mixed Models
by **Ramon C. Littell, George A. Milliken, Walter W.**
Stroup, and **Russell D. Wolfinger** . . Order No. A55235

SAS® System for Regression, Second Edition
by **Rudolf J. Freund**
and **Ramon C. Littell**. . . . . Order No. A56141

SAS® System for Statistical Graphics, First Edition
by **Michael Friendly** . . . . Order No. A56143

The SAS® Workbook and Solutions Set
(*books in this set also sold separately*)
by **Ron Cody** . . . . Order No. A55594

Selecting Statistical Techniques for Social Science
Data: A Guide for SAS® Users
by **Frank M. Andrews, Laura Klem, Patrick M. O'Malley,**
Willard L. Rodgers, Kathleen B. Welch,
and **Terrence N. Davidson** . . . . Order No. A55854

Statistical Quality Control Using the SAS® System
by **Dennis W. King**. . . . . Order No. A55232

A Step-by-Step Approach to Using the SAS® System
for Factor Analysis and Structural Equation Modeling
by **Larry Hatcher**. . . . . Order No. A55129

A Step-by-Step Approach to Using the SAS® System
for Univariate and Multivariate Statistics,
Second Edition
by **Larry Hatcher, Norm O'Rourke,**
and **Edward J. Stepanski** . . . . Order No. A58929

Step-by-Step Basic Statistics Using SAS®: Student
Guide and Exercises
(*books in this set also sold separately*)
by **Larry Hatcher**. . . . . Order No. A57541

Survival Analysis Using the SAS® System:
A Practical Guide
by **Paul D. Allison** . . . . Order No. A55233

Tuning SAS® Applications in the OS/390 and z/OS
Environments, Second Edition
by **Michael A. Raithel** . . . . Order No. A58172

Univariate and Multivariate General Linear Models:
Theory and Applications Using SAS® Software
by **Neil H. Timm**
and **Tammy A. Mieczkowski**. . . . . . . Order No. A55809

Using SAS® in Financial Research
by **Ekkehart Boehmer, John Paul Broussard,**
and **Juha-Pekka Kallunki** . . . . Order No. A57601

Using the SAS® Windowing Environment:
A Quick Tutorial
by **Larry Hatcher**. . . . . Order No. A57201

Visualizing Categorical Data
by **Michael Friendly** . . . . Order No. A56571

Web Development with SAS® by Example
by **Frederick Pratter** . . . . Order No. A58694

Your Guide to Survey Research Using the
SAS® System
by **Archer Gravely**. . . . . Order No. A55688

JMP® Books

JMP® for Basic Univariate and Multivariate Statistics:
A Step-by-Step Guide
by **Ann Lehman, Norm O'Rourke, Larry Hatcher,**
and **Edward J. Stepanski** . . . . Order No. A59814

JMP® Start Statistics, Third Edition
by **John Sall, Ann Lehman,**
and **Lee Creighton** . . . . Order No. A58166

Regression Using JMP®
by **Rudolf J. Freund, Ramon C. Littell,**
and **Lee Creighton** . . . . Order No. A58789

support.sas.com/pubs